THE
THEORY OF
EVOLUTION

·K·U·P·E·R·A·R·D·

Published in Great Britain by
Kuperard, an imprint of Bravo Ltd
59 Hutton Grove, London N12 8DS
www.kuperard.co.uk
Enquiries: office@kuperard.co.uk

Copyright © 2010 Bravo Ltd

Series Editor Geoffrey Chesler
Design Bobby Birchall

ISBN 978 1 85733 495 1

British Library Cataloguing in Publication Data
A CIP catalogue entry for this book
is available from the British Library.

Printed in Malaysia

Cover image: Computer artwork by Alfred Pasieka/Science Photo
Library

Images reproduced under Creative Commons Attribution
ShareAlike licence 3.0: page 45 © Dlloyd; page 126 © Elembis;
and under Creative Commons Attribution ShareAlike licence 2.0
France: page 122 © Rama
Image on page 145 © Wellcome Library

Simple Guides »

THE
THEORY OF
EVOLUTION

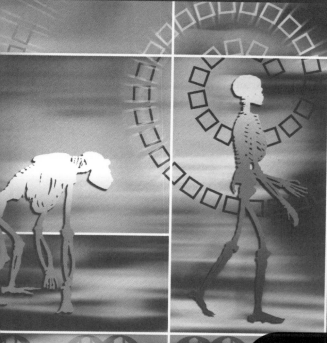

John Scotney

Contents

List of Illustrations

⊙ Evolution: Facts and Figures

3–4 billion BCE	First DNA, first bacteria
3 billion BCE	Stromatolites appear
580–540 million BCE	Cambrian Explosion – a sudden burst of life: many fossils first appear
510 million BCE	First fish
475 million BCE	First plants
390 million BCE	First insects
230 million BCE	First dinosaurs
195 million BCE	First mammals
185 million BCE	*Ichthyosaurus* thrived
145 million BCE	*Archaeopteryx* thrived
65 million BCE	End of the dinosaurs
20 million BCE	Giraffe's short-necked ancestors
6-5 million BCE	Hominins split from Apes
2.5 million BCE	*Homo habilis* in Africa
1 million BCE	*Homo erectus* in Asia
200,000 BCE	Neanderthals in Europe
190,000 BCE	*Homo sapiens* first appears in Africa
100–30,000 BCE	*Homo sapiens* in Europe. End of Neanderthals
4004 BCE	Origin of life according to Archbishop Ussher
400 BCE	Hippocrates uses Pangenesis to explain evolution
350 BCE (circa)	Aristotle teaching
1270 CE (circa)	St Thomas Aquinas writes *Summa Theologica*
1648	Archbishop Ussher publishes *Annals of the Old Testament*
1687	Newton proposes Laws of Gravitation
1754	Linneaus publishes *Genera Plantorum*
1758	Linnaeus publishes *Systema Naturae*
1794	Blake publishes 'Tyger Tyger' in *Songs of Experience*
1794	Erasmus Darwin publishes *Zoonomia*
1794	William Paley publishes *The Evidences of Christianity*
1798	Malthus publishes *An Essay on the Principle of Population*
1800	Cuvier argues that fossils represent extinct species
1802	Paley publishes *Natural Theology*
1803	Erasmus Darwin's *Temple of Nature* published posthumously
1809	Lamarck publishes *Philosophie Zoologique*
1809	Charles Darwin born
1811	Joseph and Mary Anning find *Ichthyosaurus*

1815	William Smith's geological map of Britain
1817	Darwin's mother dies
1818	Darwin goes to Shrewsbury School
1823	Mary Anning finds *Plesiosaurus*
1825	Darwin goes to Edinburgh University to study medicine
1828	Darwin enters Christ's College, Cambridge, to study theology
1829	Lamarck dies
1830	First volume of Lyell's *Principles of Geology* published
1831	Darwin graduates from Cambridge
1831–36	The voyage of the *Beagle*
1836	2 October Darwin returns to England
1836–37	Darwin writes up notes. Starts to think about evolution
1838	Darwin first formulates theory of natural selection
1839	Book later called *The Voyage of the Beagle* published
1839	Darwin marries his cousin, Emma Wedgwood
1841	Richard Owen coins name 'dinosaur'
1842	Darwin drafts brief essay with his ideas on natural selection (unpublished)
1842	Darwin publishes *The Structure and Distribution of Coral Reefs*
1844	Drafts longer essay on natural selection (unpublished)
1844	Darwin publishes *Geological Observations on Volcanic Islands*
1844	*Vestiges of the Natural History of Creation* published anonymously
1846	Darwin publishes *Geological Observations on South America*
1848	Asa Gray publishes *A Manual of the Botany of the Northern United States*
1851–54	Darwin studies and writes about barnacles and fossil barnacles
1856	Discovery of Neanderthal skeleton
1856–63	Mendel works on peas
1858	Wallace writes to Darwin outlining his idea on evolution
1858	Wallace's and Darwin's ideas presented to the Linnaean society

1859	*On The Origin of Species* published
1860	Huxley debates with Bishop Samuel Wilberforce in Oxford
1861	First *Archaeopteryx* found
1866	Mendel publishes his findings
1866	Haeckel publishes *Generelle Morphologie der Organismen*
1870	Wallace publishes *Contributions to the Theory of Natural Selection*
1871	Darwin publishes *The Descent of Man*
1880	Darwin publishes *The Power of Movement in Plants*
1881	Darwin publishes *The Formation of Vegetable Mould through the Action of Worms*
1882	Darwin dies, and is buried in Westminster Abbey
1883	Weismann introduces germ-plasm theory
1891	Discovery of *Homo erectus* (Pithecanthropus)
1894	Bateson publishes *Materials for the Study of Variation*
1900	Rediscovery of Mendel's Laws
1901	De Vries publishes *The Mutation Theory*
1903	Walter Sutton shows chromosomes carry units of heredity
1911	Morgan maps five genes on fruit fly chromosome
1918	Fisher starts work at Rothamsted
1921	Morgan maps 2,000 genes on fruit fly chromosome
1927	Peking Man discovered
1929	Lysenko emerges as leading Russian geneticist
1930	Ronald Fisher publishes *The Genetical Theory of Natural Selection*
1937	Dobzhansky publishes *Genetics and the Origin of Species*
1942	Julian Huxley publishes *Evolution, The Modern Synthesis*
1944	DNA revealed as 'transforming principle' by Oswald Avery
1949	Chargaff's Rules established
1953	Crick and Watson demonstrate structure of DNA
1961	Crick finds codons
1988	Human Genome Project begun
1996	Dolly the sheep cloned
2003	Human Genome Project completed
2006	Genome of Neanderthal man

The Voyage of the *Beagle*

HMS Beagle *in the Straits of Magellan. Mount Sarmiento in the distance*

The route taken by the Beagle, 1831-36

All Creatures Great and Small

The Lord God made them all

In 1848 Cecil Frances Alexander, the wife of an
Ulster clergyman who would later become Primate
(Chief Bishop) of all Ireland, published *Hymns for
Little Children*. It contained 'Once in Royal David's
City' and 'There is a Green Hill Far Away', but the
most famous of her hymns began:

*All things bright and beautiful, all creatures
    great and small,*
*All things wise and wonderful, the Lord God
    made them all.*
*Each little flower that opens, each little bird
    that sings,*
*He made their glowing colours; He made
    their tiny wings.*

Her charming and optimistic words in fact reflect
the view of the majority of scientists in the 1840s.
While not necessarily believing that God had
completed the job in seven days, they did think that
each species of plant and animal had been individually
crafted in a distinct act of creation. The complexity of
the living world certainly suggests that such a task
would have required an omnipotent, all-wise Creator.

Recent estimates give a total of 287,655 plant species. These are the known ones. Some years ago I made a radio programme at the Royal Botanical Gardens, Kew. One of the senior scientists had worked in the Brazilian rain forest, where a single acre can contain 750 types of tree and 1,500 species of higher plant life. She told me that a very small proportion of forest plants had been identified, and that there were countless unrecorded varieties.

⊙ *Creation of the Birds and Fishes. Etching by Wenceslaus Hollar, 17th century*

According to the distinguished naturalist David Attenborough, a similar discrepancy is true of the insects in the rain forests. As it is, leaving aside the crowded world of bacteria, over 1,250,000 species of animal are on record, 950,000 of them insects. By contrast, a mere 5,416 species are mammals.

Why are there so many species? Why are there so many apparently minor variations between similar animals and plants? Over a third of insect species are miscellaneous types of beetle, which led the geneticist J. B. S. Haldane to remark, 'If one could conclude as to the nature of the Creator from a study of his creation it would appear that God has a special fondness for stars and beetles.'

Haldane also warned – and this gives pause to any scientific theorist – that the world is not only stranger than we imagine, but also stranger 'than we *can* imagine'. Among strange creatures that we do know of are the male barnacles, studied by Darwin, that spend their whole life inside the female. Other, more familiar, quaint creations are stick insects, which make themselves look exactly like twigs; and the puffer fish that blows itself up to make itself hard to swallow, and contains poison to kill you off if you do manage to swallow it – neither of which stops Japanese gourmets eating it as an expensive delicacy. The cunning defences of puffer fish and stick insects are matched by the eagle's eyesight and claws and the anteater's long nose, which are perfectly adapted to catching their chosen prey.

The battle for survival

Behind this vast, ingenious, spendthrift diversity every creature is simply struggling to eat and not be eaten, and to reproduce. Not just individual creatures: whole species are struggling to survive. Many thousands of them have failed to make it, while others have battled on from almost the very beginning. The stromatolites of Shark Bay, Australia, are life forms that have been around virtually unchanged for three billion years, while the recently discovered 'spiral poo worm' that lives deep in the ocean bed has excited scientists because its spiral-shaped faeces are identical to fossil faeces dating back hundreds of millions of years.

Lords of Creation and little worms

We humans are latecomers: in the biblical account we arrive at the end of the last day before God rested, and palaeontologists introduce *Homo sapiens* in the last 100,000 years or so of the multi-billion year history of living things. Does this mean that all the rest of Creation was leading up to us, its ultimate triumph? That was what the early evolutionists believed. But are we truly the Lords of Creation? For millions of years the mighty dinosaurs were exactly that, until one day a rogue asteroid crashed into the Earth. If another asteroid were to strike tomorrow that would be the end of us, and all our works, while the humble spiral poo worms would probably live on, secure in the ocean depths, and carry on, delivering their spiral faeces, regardless. Life is wonderful, detailed, complex and interrelated, and at the same time wasteful, cruel and contradictory.

The entangled bank and the tiger

In a celebrated passage near the end of *The Origin of Species* Charles Darwin wrote about the kind of ordinary, overgrown bank you might find at the end of a garden, or beside a railway track, ignored as the trains speed by:

'It is interesting to contemplate an entangled bank, clothed with many plants of many kinds, with birds singing on the bushes, with various insects flitting about, and with worms crawling through the damp earth, and to reflect that these elaborately constructed forms, so different from each other, and dependent on each other in so complex a manner, have all been produced by laws acting around us.'

Or, alternatively, we can look at the savage, beautiful Tiger and ask, as did William Blake, the half-mad genius, whose life overlapped with Darwin's for eighteen years:

When the stars threw down their spears,
And watered heaven with their tears,
Did he smile his work to see?
Did he who made the Lamb, make thee?

Tyger! Tyger! Burning bright
In the forests of the night,
What immortal hand or eye
Dare frame thy fearful symmetry?

Blake was a profoundly believing if eccentric Christian – yet he could not but question how a God who saw himself as the 'Prince of Peace' and identified himself with the Holy Lamb could have created such an elegant, effective and cruel killing machine as the Tiger.

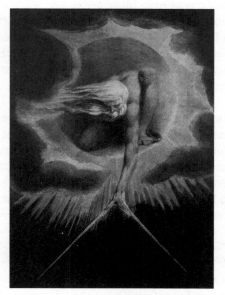

⊙ *The Ancient of Days. Watercolour relief*
etching by William Blake, 1794

Chapter 1

Young Darwin and God's Unchanging Creation

The Devil's chaplain

The same issue worried Charles Darwin: how could a kindly Creator be responsible for so much cruelty? 'What a book the devil's chaplain might write on the clumsy, wasteful, blundering, low, and horribly cruel work of nature!' he commented to his friend Joseph Hooker in 1856, well before he published *The Origin of Species*; and in a letter to the Harvard Professor Asa Gray in 1860, the year after he had published it:

> 'There seems to be too much misery in the world. I cannot persuade myself that a beneficent and omnipotent God would have designedly created the Ichneumonidae [parasitical wasps] with the express intention of their feeding within the living bodies of caterpillars, or that a cat should play with mice.'

The Great Chain of Being

Darwin must have been familiar with the philosophical justification for the existence of cruelty and suffering that dates back to Aristotle

(384–322 BCE). Everything, it was said, is an essential part of the divine scheme of things, hence nothing is bad, only imperfect, less good. This explanation of the complexity and interdependence of every aspect of life, of what was called 'Creation', came to be known as 'The Great Chain of Being'. According to St Thomas Aquinas (1224–74) everything fitted together like an enormous jigsaw puzzle, and had done so since the six days when God made it all. Fitting everything together had required a Supreme Intelligence, and that was what God was.

Aquinas postulated a ladder descending from God, through archangels, then angels, mankind, the animal kingdom, plants and minerals, down to mud and finally sheer nothingness. Other great philosophers, notably Descartes (1595–1650) and Spinoza (1632–77), adapted and developed the idea. As Descartes saw it, all 'Creation' is inextricably, mutually interdependent and is, and must be, 'perfect' (that is to say, complete). Hence it must contain every possible kind of creature, even the imperfect and downright nasty ones. God had to make Ichneumonidae, or he would have left out a piece of the jigsaw. It is not so remote from modern concepts of 'biomass' and 'biosphere'.

More alien to modern thinking is the suggestion that because Creation was hierarchic, with angels at the top and mud at the bottom, everything and everyone had a proper place. Or, as 'All Things Bright and Beautiful' puts it in a verse that today is usually left out:

The rich man in his castle, the poor man at his gate,
God made them, high or lowly, and ordered their estate.

In an age of rapid change we can easily forget that our ancestors saw the world as essentially static and stable; attempts to change the fixed order were dangerous. The Universe remained in harmony so long as everything and everyone stayed where they were put and didn't rock the boat. In *Troilus and Cressida* Shakespeare makes Ulysses tell how the stars and planets 'observe degree, priority and place':

'Take but degree away, untune that string,
And, hark, what discord follows! Each thing meets
In mere *oppugnancy* ... '

An interesting word that: 'oppugnancy'. Shakespeare says that the alternative is everything fighting everything else, or perhaps 'struggling against everything else'. In the late eighteenth century the English economist Thomas Malthus suggested that this was precisely how population was regulated. Everybody is in competition for food and living space, and because we breed too many children for the Earth's limited resources the weakest fail to survive. Malthus died in 1834, when Charles Darwin was twenty-five years old and half a world away; but his ideas were to prove something of a revelation to the young scientist.

A disgrace to the family

Charles Darwin was the grandson of the famous physician, philosopher, poet and eccentric, Erasmus Darwin. Erasmus had been a well-known doctor, but not an especially rich one. In 1786, he brought his twenty-year-old son Robert, newly qualified as a doctor, from Derby to Shrewsbury, and left him with £20 in his pocket to set up a medical practice. Robert became one of the richest and most influential men in the county of Shropshire. He was a good doctor, and also a shrewd businessman and property speculator, and in 1796 married Susannah Wedgwood, the daughter of one of Erasmus's closest friends, Josiah Wedgwood, the great pottery manufacturer. Charles was the fifth of Robert and Susannah's six children, and Susannah died in 1818 when he was nine years old. Robert had no great hopes for his son:

'To my deep mortification my father once said to me, "You care for nothing but shooting, dogs, and

rat-catching, and you will be a disgrace to
yourself and all your family."'

Shrewsbury and Edinburgh

Charles was sent to Shrewsbury School as a
boarder, even though it was scarcely a mile from
the family home. His father had studied medicine
at Edinburgh University, for which he had great
respect, and Charles was expected to follow the
family tradition and become a doctor:

'As I was doing no good at school, my father
wisely took me away at a rather earlier age than
usual, and sent me (October 1825) to Edinburgh
University, where I stayed for two years ... But
soon after this period I became convinced that my
father would leave me property enough to subsist
on with some comfort, though I never imagined
that I should be so rich a man as I am; but my
belief was sufficient to check any strenuous
efforts to learn medicine.'

There was more to it than that. He had to
watch two operations in those brutal days before
anaesthetics: one was on a child, and the soft-
hearted Charles fled in horror. He was clearly not cut
out to be a doctor. However he met some interesting
people in Edinburgh, among them Dr Robert Grant,
who introduced him to Lamarck's theory of evolution,

and a 'very pleasant and intelligent' Guyanese ex-slave, John Edmonstone, who taught him taxidermy – the art of stuffing animals.

Cambridge

Neither Erasmus nor Robert Darwin was an orthodox believer, but every rich man knew that the thing to do with a dim younger son was to get him into the Church and fix him up with a cosy living as a country parson. So Charles was sent to Cambridge to study theology. He missed the first term because he had forgotten all his Classics from school and had to take a crash refresher course in Greek and Latin.

'During the three years which I spent at Cambridge my time was wasted, as far as the academical studies were concerned, as completely as at Edinburgh and at school.'

So Charles himself admits in his brief autobiography – though it hasn't stopped his old college putting up a statue to him. He also admits to preferring shooting to study: he was a first-rate shot, and spent far too much time out with his guns at Maer in Staffordshire, the estate of his 'Uncle Jos' (Wedgwood). But, he adds, significantly, 'no pursuit at Cambridge was followed with nearly so much eagerness or gave me so much pleasure as collecting beetles.' Years later he could remember the exact circumstances in which he had found many of his specimens, and recalled his delight at seeing his name in Stephens' *Illustrations of British Insects*. Imagine, then, the thrill he felt when only a few years later in Brazil he was to discover sixty-eight new species of beetle in a single day.

Henslow

For all his apparent idleness there must have been something special about the young Darwin, for both at Edinburgh and at Cambridge some of the most perceptive scientific minds welcomed his company and treated the undergraduate almost as an equal. At Cambridge he attended the public scientific lectures of John Stephens Henslow, who combined being a very orthodox Anglican clergyman with the professorship of botany. Henslow kept open house once a week. Darwin became a regular attender, and they got on so well that during his last year he and Henslow often went on long walks together in the countryside. But Darwin still had to get his theology degree, and as his finals approached he worked feverishly to catch up on the studying he had missed while out shooting, collecting beetles, or walking with Henslow. One of his set books was Dean Paley's *Evidences of Christianity*. Darwin was struck by Paley's logical, clear, intelligent analysis, and went on to read his *Natural Theology*. The two books made a great impression on him.

The Watch on the Heath

William Paley was the leading Christian thinker of his age. His approach was in a sense scientific, since he did not argue from authority, as in, 'It must be true because the Bible says so'. He based his case instead on observation and logic. He is most famous for his concept of the 'Watch on the Heath'. The key passage states:

'In crossing a heath, suppose I pitched my foot against a **stone** and were asked how the stone came to be there, I might possibly answer that for anything I knew to the contrary it had lain there forever. But suppose I had found a **watch** upon the ground, and it should be inquired how the watch happened to be in that place, I should hardly think of the answer which I had before given, that for anything I knew the watch might have always been there. Yet why should not this answer serve for the watch as well as for the stone? Why is it not as admissible in the second case as in the first? For this reason, and for no other, namely, that when we come to inspect the watch, we perceive – what we could not discover in the stone – that its several parts are framed and put together *for a purpose*, that, if the different parts had been differently shaped from what they are, or placed after any other manner or in any other order than that in which they are placed, no motion at all would have been carried on in the machine ...'

It is a sophisticated development of the 'Great Chain of Being'. Paley suggests that the Universe resembles a watch: it is complex, it fits together with great ingenuity, it clearly has a purpose, as does the watch, and is so brilliantly organised that if you change one bit it will stop working. Surely the odds against this amazing machine having come about by chance are impossibly colossal? It must have been designed. Nowadays this theory is called 'intelligent design', but philosophers have always referred to it as the 'teleological argument', from the Greek word *telos*, meaning 'purpose'. As Aquinas expressed it: how could the blind forces of nature all work together to sustain life and enable progress unless there were some intelligent mind directing them? Paley similarly asked how birds knew where to migrate to, or how they would instinctively build nests when they had never done so before, and were no more capable of thought than a watch was, unless some Supreme Mind had designed them to do just that. How could the complex structure of the eye come about by chance?

This particular problem was to concern Darwin for years, and he reports that, as a student, he considered Paley's arguments 'conclusive'. In a more subtle form they remain the convinced believers' best intellectual case against what they see as the 'blind chance' of Darwinism.

Grandfather Erasmus

But even in Paley's day there were alternative theories. Indeed, he wrote his books to refute them. The chief proponents of the alternative view were Jean-Baptiste Lamarck, who will be considered in more detail in a later chapter, and, more especially, Charles's own grandfather, Erasmus Darwin.

Erasmus was one of those larger-than-life eighteenth-century figures like Dr Johnson. In fact he came from the same Midlands town – Lichfield – as Johnson, and resembled him physically. But Erasmus was as radical in his views as Johnson was conservative.

Erasmus founded the Lunar Society of Birmingham – a group of brilliant men who met every month at the full moon, not for its mystic resonance, but simply because the bright

⊙ *Erasmus Darwin, by Joseph Wright of Derby, 1792*

moonlight made it easier to ride home. The other main members were James Watt, Joseph Priestley, Matthew Boulton and Josiah Wedgwood (Charles

Darwin's other grandfather): the four greatest figures of the Industrial Revolution.

Erasmus Darwin was a fascinating man, and more about him can be found in any of a number of books by Desmond King-Hele. He, like his son Robert, was a Unitarian – he didn't believe in the Holy Trinity, seeing Jesus as simply a remarkable human being, and was probably secretly an atheist. This makes Charles's decision to become a Church of England vicar all the more curious.

'First forms minute'

Erasmus attracted Paley's hostility by his two vast works *Zoonomia* (1794) and the posthumously published *The Temple of Nature* (1803). They both anticipated Lamarck's ideas. Here is Charles Darwin writing to Thomas Huxley:

> '[It is] curious to observe how exactly & accurately my Grandfather (in Zoonomia Vol. I. p. 504 1794) gives Lamarck's theory. I will quote one sentence. Speaking of Birds Beaks, he says "All of which seem to have been gradually produced during many generations by the perpetual endeavour of the creatures to supply the want of food, & to have been delivered to their posterity with constant improvement of them for the purposes required."

Charles then points out that Lamarck did not publish until 1809, fifteen years later. He doesn't mention that Grandfather Erasmus had also anticipated Charles's own ideas based on his study of finches' beaks!

The Temple of Nature is written in verse. Here are a couple of extracts from a famous passage in which Erasmus accurately postulated the evolution of plant and animal life from sub-microscopic spores in the sea to mankind itself:

ORGANIC LIFE beneath the shoreless waves
Was born and nurs'd in Ocean's pearly caves;
First forms minute, unseen by spheric glass,
Move on the mud, or pierce the watery mass;
These, as successive generations bloom
New powers acquire, and larger limbs assume
Whence countless groups of vegetation spring,
And breathing realms of fin, and feet,
and wing...

Imperious man, who rules the bestial crowd,
Of language, reason, and reflection proud,
With brow erect who scorns this earthly sod,
And styles himself the image of his God;
Arose from rudiments of form and sense,
An embryon *point or* microscopic ens!

(*Embryon* is the correct Greek form for what we now call an embryo, and *ens* is Latin for what we now call an entity.)

No wonder Paley felt the need to take up his pen!

The speculator and the scientist

Darwin admitted that his grandfather's ideas had probably influenced his own, since as a young man he had greatly admired *Zoonomia*. However, when he came to re-read it, 'I was much disappointed; the proportion of speculation being so large to the facts given.' Paley's ideas, by contrast, seemed more logically organised and factually based.

Erasmus had a brilliant speculative mind, capable of making giant leaps of imagination. Charles had the ability to collect a vast array of facts, sort them, categorise them, throw out ideas that didn't fit the facts, and ask himself what conclusion the facts led to. As he put it: 'I am a sort of machine for observing facts & grinding out conclusions'. Erasmus thought like an eighteenth-century savant; Charles had the mind of a scientist, which was why it was Charles, and not his grandfather, who revolutionised scientists' understanding of the living world.

Time to kill

In 1831 Charles got his theology degree: he wasn't considered bright or hardworking enough to go for Honours, but among those lesser students who sat for pass degrees he proved to be one of the better ones, chiefly thanks to his knowledge of, and sympathy with, the ideas of Dean Paley. Because he had come up to Cambridge three months late, he had to stay on two extra terms, and Henslow arranged for him to go on a geological investigation in North Wales with Professor Sedgwick, the famous geologist after whom the Sedgwick Museum at Cambridge was named. Though Charles left as soon as the shooting season started, the expedition taught him a vital lesson:

'Nothing before had ever made me thoroughly realise, though I had read various scientific books, that science consists in grouping facts so that general laws or conclusions may be drawn from them'

For five years of his young life he would assiduously collect facts without really knowing where they would take him or what conclusions he would draw.

Chapter 2

The Voyage of the *Beagle*

Charles now had his degree in theology, or divinity, as it is called in Cambridge. His piety was genuine, and he could quote scripture by the ream. All he had to do now was to wait for a suitable parish to turn up. His increasing interest in insects, plants, animals in general, fossils and geology were perfectly normal for an English vicar at the time; Henslow himself was a clergyman. A parish would cost money to secure, but his father was rich and his uncle Josiah Wedgwood would probably also help, being after all one of England's leading industrialists. As it happened, 'Uncle Jos' helped in a quite different way. Charles wanted to travel a little before he settled down – the 'gap year' is by no means a modern invention. Unhappily the friend he planned to travel with died suddenly.

Gentleman's dining companion

Then, wholly unexpectedly, a letter arrived from Henslow. A Captain Robert Fitzroy was to command a Royal Navy brig called the *Beagle* sailing to survey the southern part of South America, and at the same time testing the new naval chronometers as a way of assessing longitude. The survey was planned to last two years. We tend to forget it, but this was an important voyage irrespective of Darwin's involvement, and Fitzroy himself was a significant scientific innovator. He was twenty-six at the time, the grandson of the Duke of Grafton, and a comfortably rich young man. So much so that he was having the *Beagle* refitted at his own cost. But it was still only a brig, and therefore not large: brigs were the smallest class of fighting vessel in the navy, and this one was 90 feet (30 metres) by 24 feet (8 metres), though it did carry ten guns. Darwin later claimed that he was taken on as unpaid 'ship's naturalist'. This was not wholly true: the *Beagle* already had a naturalist in the person of its surgeon, Robert McCormick. The post offered was that of 'gentleman's dining companion' – someone for Fitzroy to talk to, someone who was a gentleman, and who had wide scientific interests.

At first Robert Darwin said no, he had already paid out a fortune for his far from thrifty son during his five years at various universities. Then Uncle Jos intervened with one of those typically avuncular letters on the lines of 'the trip will be the making of the lad'. Robert had a lot of respect for

Josiah, and acquiesced in the plan despite being convinced that Charles would somehow manage to overspend while he was away.

Uncle Josiah took Charles down to London in his carriage to meet Captain Fitzroy. Fitzroy took one look at Darwin and told him, sorry, no, he had changed his mind. Fitzroy later explained that Darwin's nose had condemned him. He had a theory that he could judge people by their noses, and Darwin's was unacceptable. Fortunately Fitzroy's friend, who had agreed to go in his stead and was no doubt agreeably nosed, suddenly cried off, perhaps when he saw the size of the *Beagle*. By chance Darwin looked in on Fitzroy five minutes later and got the job, nose notwithstanding.

After a three-month delay, the *Beagle* finally sailed on 27 December, 1831, with the twenty-two-year-old Charles Darwin aboard. He had never previously travelled further afield than Scotland and Wales. He was very seasick.

A world of wonders

It was to be an incredible experience lasting not two years but five, though only about eighteen months were actually spent at sea. All the while Darwin observed and noted. The scientifically inclined young wastrel grew and grew in intellectual stature, maturity and understanding and, despite continued acute seasickness, he loved it. The first landfall was the Cape Verde Islands:

'The scene, as beheld through the hazy
atmosphere of this climate, is one of great
interest; if, indeed, a person, fresh from sea, and
who has just walked, for the first time, in a grove
of cocoa-nut trees, can be a judge of anything but
his own happiness'

Among the books he had brought with him was
Professor Lyell's groundbreaking work on geology,
published the previous year (see page 49). Darwin
put it to use right away at this, his first landfall:

'Upon examination this white stratum is found
to consist of calcareous matter with numerous
shells embedded, most or all of which now exist
on the neighbouring coast. It rests on ancient
volcanic rocks, and has been covered by a
stream of basalt, which must have entered the
sea when the white shelly bed was lying at
the bottom …'

Everything, not just the geology, seems to come
together on the voyage. The beetle collector finds a
paradise of beetles in Brazil; the crack shot is able
to bring down vital specimens, from tiny finches
to a condor with wings eight and a half feet across.
The time he spent in Edinburgh learning to stuff
animals proves invaluable.

He sees mocking-birds, flightless rheas and
vampire bats; he is puzzled by a platypus and a
bandicoot. In a remote village in Uruguay he finds

boys using a huge fossil skull for
target practice. He sees a volcano in
action in Chile and speculates on the
cause of earthquakes; in Tahiti and
Australia he sees coral reefs and collects
evidence that will later lead him to
contradict the theories of the great
Lyell. And always he observes and
records and measures.

⊙ Rhea darwinii, *painted by John Gould*

'Looking backwards, I can now
perceive how my love for science gradually
preponderated over every other taste. During
the first two years my old passion for shooting
survived in nearly full force, and I shot myself
all the birds and animals for my collection; but
gradually I gave up my gun more and more, and
finally altogether, to my servant, as shooting
interfered with my work, more especially with
making out the geological structure of a country.
I discovered, though unconsciously and insensibly,
that the pleasure of observing and reasoning was
a much higher one than that of skill and sport.'

Over the years Darwin sent back specimens and
letters to England, particularly to Henslow, in
Cambridge, who considered the letters of sufficient
scientific interest to print them in a pamphlet, which
he sent to other leading experts. Darwin's name had
become well known in these circles long before he
got back home.

People

But the sporting, sociable young man had not transmogrified into a narrow-minded pedant. Darwin was deeply interested in the people he met, even if they sometimes disturbed him. In Brazil he was frightened by the young boys' ability to throw knives, in New Zealand he and his party got into a firefight with Maoris, and was hardly comforted by being told that they were not cannibals, at least not recently. On the rain-swept Falkland Islands a lone British officer was in charge of a population of 'runaway rebels and murderers'. In Tierra del Fuego, the utter primitiveness of the naked inhabitants astounded him. By contrast, the nobility of the Tahitians moved him, and the parson-to-be was deeply affected by their simple piety:

'Before we laid ourselves down to sleep, the elder Tahitian fell on his knees, and with closed eyes repeated a long prayer in his native tongue. He prayed, as a Christian should do, with fitting reverence, and without the fear of ridicule or any ostentation of piety. At our meals neither of the men would taste food, without saying beforehand a short grace.'

But it was the cowboys – the gauchos on the Argentinean Pampas – he liked best. He admired their stupendous riding and their amazing ability to swing a *boleadoras* at full gallop. He tried it himself:

'One day, as I was amusing myself by galloping and whirling the balls round my head, by accident the free one struck a bush; and its revolving motion being thus destroyed, it immediately fell to the ground, and like magic caught one hind leg of my horse; The Gauchos roared with laughter; they cried they had seen every sort of animal caught, but had never before seen a man caught by himself.'

Darwin was to be cooped up with Robert Fitzroy for long periods in a ship not much bigger than a couple of buses. It does them both credit that their mutual respect survived, though they had terrible rows. Darwin hated slavery; his grandfather Josiah's factory had produced a famous porcelain medallion of a kneeling black slave in chains and the words, 'Am I not a man and a brother?' But let Darwin take up the story:

'Early in the voyage at Bahia, in Brazil, he [Fitzroy] defended and praised slavery, which I abominated, and told me that he had just visited a great slave-owner, who had called up many of his slaves and asked them whether they were happy, and whether they wished to be free and all answered "No." I then asked him, perhaps with a sneer, whether he thought that the answer of slaves in the presence of their master was worth anything?'

Fitzroy cracked, and after a screaming tantrum told Darwin he could no longer bear having him on board – but within hours had calmed down and invited him back. In fact Fitzroy had a markedly unstable temperament and the subtext of Darwin's job was that he was always on suicide watch. There was a history of depression in Fitzroy's family: his uncle Viscount Castlereagh had killed himself, and Robert himself was ultimately to do the same in 1865. One reason he wanted a companion on board was that the *Beagle*'s previous commander had committed suicide. Command at sea was a lonely job in those days, even if you were not a depressive like Fitzroy.

Fortunately he and Darwin had plenty to occupy themselves and worked hard to get on well. Fitzroy was a religious fundamentalist, while Darwin followed Dean Paley. But the would-be vicar was beginning to notice things that struck him as curiously ambiguous.

Did He who made the lamb ...

There were the fossils he collected, for instance. In Argentina he found enormous ground-sloths, the Megatheres, and the colossal, armour-plated glyptodont. Could they be related to the armadillos he had observed elsewhere in South America? And what was he to make of the primitive, seafood-eating platypus he would see in Australia? It has a huge, duck-like beak, a body covered in fur, four legs, mammary glands, poisonous spurs, very low blood temperature and the flat tail of a beaver. And it lays eggs. The seashells he found on 13,000-foot (3,962 m) high mountains near Valparaiso also posed questions. But, above all, there were the Galapagos Islands.

Nothing could be less inviting ...

On 15 September, 1835, nearly four years into the voyage, they sighted the Galapagos Islands rising from the Pacific Ocean, six hundred miles off the coast of Ecuador. Two days later Darwin landed on Chatham Island:

> 'Nothing could be less inviting than the first appearance. A broken field of black basaltic lava, thrown into the most rugged waves, and crossed by great fissures, is everywhere covered by stunted, sun-burnt brushwood'

Technically the islands belonged to Ecuador, but they were inhabited mainly by European whalers and had an English vice-governor. Darwin spent five weeks there, and analysing what he had seen would be pivotal to his later thinking.

The animals and plants were not startlingly unusual. When Darwin was to see the marsupials of Australia he would remark that: 'An unbeliever might exclaim, "Surely two distinct Creators must have been at work!"' By contrast, what he saw on the Galapagos Islands was familiar but different. He had seen the dragon-like iguanas before, but they lived in forests and ate fruit, leaves and the occasional egg. Here on one island they had developed massive claws, which enabled them to climb rocky cliff-faces, and they ate seaweed. He was familiar with tortoises, but here they were enormous, so big you could ride on them; he did so, and fell off. Depending upon which island they came from, the tortoises differed slightly. Some of the islands, like Chatham Island, were dry and lacked ground vegetation, and here the tortoises had longer necks and the top shell had turned back to allow them to reach up for scrappy leaves on the brushwood; elsewhere, on greener islands, the ground feeders had shorter necks and downturned shells.

The birds, too, were familiar, but different: there were cormorants of a sort, but they could not fly. Most interesting of all were the finches and mocking-birds Darwin shot. He mused how otherwise similar birds had wholly different types of beak, each adapted to their particular diet on the different islands. He wrote:

'The most curious fact is the perfect gradation in the size of the beaks in the different species of Geospiza [finches], from one as large as that of a hawfinch to that of a chaffinch, and (if Mr. Gould is right in including his sub-group, Certhidea, in the main group) even to that of a warbler … there are no less than six species with insensibly graduated beaks.'

Teamwork

Modern scientific research is carried out by teams.
It is pleasant to contemplate a heroic age of science
when giants like Darwin bestrode the natural world
and single-handedly made earth-shattering
discoveries. But note the name 'Mr Gould', above.
Even Darwin would have not have got far without
the help and inspiration of others. He freely admits
he could not see any difference between the
various tortoises. Nicholas Lawson, the English
Vice-Governor of the islands, who presumably had
plenty of opportunity to observe them, pointed out
the differences, as well as those in other animals.
As for 'Darwin's finches', he did not even know that

Darwin's finches, or Galapagos finches, painted by John Gould

they were all finches, but thought some of them were other birds, such as blackbirds. When he got back he donated the whole lot to the Royal Zoological society, where John Gould, the leading British ornithologist, immediately spotted their significance. They were finches, they were closely related, yet there were twelve distinct species in the ten or so islands and, most significantly, the different types of finch with differing beaks appeared on different islands, each with a different source of food.

Covington's finches

While praising Darwin's methodical approach in general, it has to be admitted that he forgot to label each specimen of finch with the island it came from, which would have made Gould's careful analysis impossible. Luckily Darwin's servant, Syms Covington, had made his own finch collection and had carefully noted the island of origin. In practice it was mainly these birds that Gould examined and of which Darwin wrote:

'It was evident that such facts as these, as well as many others, could only be explained on the supposition that species gradually become modified.'

Gould meanwhile remained a creationist all his life, and Covington went off to Australia to run a public house.

To confuse matters even more, although Darwin himself refers to finches, and 'Darwin's finches' have become famous, it was actually the differences between the mocking-birds on Chatham Island and those on the other Galapagos Islands that first led him to question the stability of species.

Linnaeus and taxonomy

In the light of Darwin's work on the Galapagos Islands, it is worth considering again his sudden realisation that, 'Science consists in grouping facts so that general laws or conclusions may be drawn from them'. This unromantic, almost bureaucratic task of sorting things into categories is at the heart of scientific method. Without the work of a Swedish botanist who had died seventy years earlier Darwin would not have known where to begin.

In his two great volumes *Genera Plantorum* (1754) and *Systema Naturae* (1758), Carol Linnaeus had classified all nature within three 'Kingdoms' – Animal, Mineral and Vegetable. Each kingdom was broken down into 'Classes'; these were broken down into 'Orders', which in turn were divided in 'Genera' (singular: genus) and these into 'Species'. Linneaus called his system 'taxonomy', from the Greek *taxis*, arrangement, and *nomos*, name – an arrangement of names.

The actual arrangement has changed greatly over the years, but the principle remains the same and has proved to be an immensely valuable tool, enabling scientists to classify living organisms and establish their relationship with each other, which is basically what Darwin did. Without Linnaeus's 'Species' there could have been no *Origin of Species*.

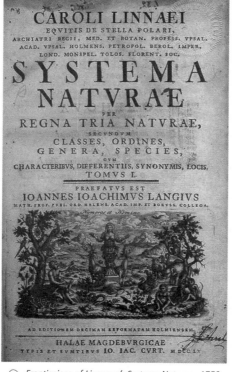

⊙ *Frontispiece of Linnaeus's* Systema Naturae, *1758*

Chapter 3

The Fossil Record

All creatures great ...

The fossils Darwin found in South America would impress Richard Owen, the leading palaeontologist and the man who coined the word 'dinosaur' ('terrible lizard'). Owen helped Darwin to classify them. To Owen, each new class or species was a separate creation, based on 'archetypes' in God's mind and launched according to God's celestial calendar. But the more Darwin considered the fossils, the more he found this explanation hard to accept: the colossal sloths may have been a big as a bus, but they were clearly related to their modern counterparts. It was puzzling.

... and small

Fossils had puzzled people for millennia. Like Darwin the Ancient Greeks had wondered why what were obviously seashells appeared inland and up mountains. Fifteen hundred years later, in 1000 CE, the great Arab thinker Avicenna proposed that the layered strata of the Earth had been caused by the ebb and flow of great seas – which does indeed explain the origin of the sedimentary rocks in which most fossils are found. In the seventeenth century Athanasius Kircher, the father of Egyptology, and, incidentally the inventor of the megaphone, was convinced that the huge bones that occasionally turned up had belonged to giant men. But the big bones did not trouble people so much as did the smaller, strange, curled creatures the Roman philospher Pliny had named 'horns of Ammon', and which later came to be called 'ammonites' – they were quite simply like nothing on Earth. Had they once lived here and then died out? Had God, who created all things, made a mistake and written it off?

Snakestones

When Darwin was sailing the world the significant developments in the understanding of fossils either had occurred only a few years earlier or were still happening. A key figure was just ten years older than he was, but from a very different background.

⊙ *Ammonite fossil from Lyme Regis, Dorset*

Mary Anning was a Dorset girl from the seaside town of Lyme Regis, where the outcrops of Blue Lias, a sedimentary rock from the Jurassic period,

⊙ *Portrait of Mary Anning, British Museum*

were packed with fossils. Mary's father, Richard Anning, was a not very successful furniture maker, and to supplement his small income he sold the curiosities he found on the beach, having first carefully cleaned them up and polished them. Particularly valuable were the spiral stones called 'snakestones' – ammonites – which supposedly had magic healing power. In 1807 Richard fell off a cliff. He never really recovered and died three years later, deeply in debt. The Anning family went 'on the parish', which gave them three shillings a week to live on. Mary, aged eleven, decided to restore the family business; like her father, she collected fossils.

A small fortune

In 1811, soon after she had started, her brother found an odd-looking fossil in rock that had fallen from the cliff. It resembled the head of a lizard; with one difference. It was 4 feet (120 cm) long. Together they found and dug out the rest of it. It looked like a monstrous hybrid of a crocodile and a dolphin, and they sold it for £23, a small fortune, enough to enable the family to live comfortably for six months. It came to be called an *Ichthyosaurus*, or 'fish lizard', though it turned out to be neither fish nor lizard. Mary was in business. She was twenty-three when she found the first *Plesiosaurus* ('nearly-lizard' – they were more cautious in their naming of this one).

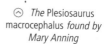

⊙ *The* Plesiosaurus macrocephalus *found by Mary Anning*

Mary learned French to study the works of the greatest living expert, Georges Cuvier; she learned anatomy; she first correctly identified coprolites (fossilised dinosaur dung); and when she died, at the young age of forty-six, she was given a grand eulogy by the London Geological Society, even though the Society wouldn't admit even upper-class women as members.

The Great Flood

The word 'evolution' suggests things evolving – simple, single-celled life forms becoming multi-celled, developing into either plants or animals; life growing in the sea and then a bold fish wriggling ashore – and the rest is history, or prehistory. This was the theory of 'transmutation', Erasmus Darwin's belief that we humans are the end of a long process of continual improvement. But it was clear that the creatures that Mary Anning and others dug up were not simple prototypes, but at least as complex as modern animals – so perhaps, after all, God had created them at the same time as the other animals. But why had they become extinct? Had they been swept away in Noah's Flood? Certainly when Mary Anning started collecting ammonites, there were leading geologists who believed the structure of the Earth's surface could be explained by its having once been entirely covered in water during the forty days of the biblical flood.

⊙ *Duria Antiquior. Watercolour by Henry De la Beche depicting early life forms in Dorset based on fossils found by Mary Anning*

Neptune, Pluto, Catastrophes and Charles Lyell

The experts who promoted the flooded world theory were called Neptunists, after the Roman god of the sea. But Charles Lyell, the greatest living geologist, whose first volume on geology accompanied Darwin round the world, was not one of them. Their rivals were called Plutonists, after the Roman god of the underworld, which Victorians identified with Hell – they attributed everything to fiery volcanoes. The leading Plutonist was the Scot James Hutton, who postulated the slow hardening of the molten rock over many millions of years. Lyell wasn't a Plutonist either. Rather, Lyell, arguably the founder of modern Geology, agreed with them both. In

⊙ *Charles Lyell*

his *Principles of Geology*, published in 1830 – the book Darwin took round the world – Lyell affirmed that the Earth changes very gradually, is immeasurably old, and that the strata or layers of rocks could be laid down by seas depositing silt (sedimentary rocks) or by volcanic activity (igneous rocks). He did not accept Georges Cuvier's 'Catastrophe' theory, however. Cuvier believed change had come in a series of catastrophic jolts – one of them being the Flood. Modern paleontologists, though, endorse the role of catastrophes: about 96 percent of species were killed off 250 million years ago as a result of some disaster, and the dinosaurs

were probably destroyed by an asteroid hitting the
Earth 65 million years ago. Above all, what Lyell
proved was that the Earth was very, very old.
This was to be a key factor in Darwin's hypothesis.

Cuvier, Smith, Sedgwick and strata

Mary Anning had learned French in order to to read
Cuvier because Georges Cuvier was *the* expert on the
study of fossils, or 'palaeontology', as it was coming to
be called. Cuvier demonstated beyond all reasonable
doubt that many, many species had disappeared,
become extinct, over the milllennia. At the same time
Cuvier and the Englishman William Smith developed
'stratigraphy', or the study and mapping of the different
strata of rocks that can be seen exposed in many cliffs
or layered on a gigantic scale in the Grand Canyon.
Smith saw the different types of fossils in the various
layers as a way of distinguishing between strata, but
Cuvier came at it from the other direction and worked
out that the relative age of fossils can be calculated by
looking a the layer of rock in which they are lodged and
its distance from the current surface of the planet.

Rock strata and mountains began to be catalogued
into 'ages' and given apropriate names. In the 1830s
Sedgwick mapped the 'Cambrian' rocks while Lyell
worked on the 'Tertiary' age. Eventually a whole
chronology emerged. Most of this happened when
Darwin was in his twenties and thirties. Meanwhile
the 'comparative anatomists' (which is what Cuvier
really was), were beginning to observe how species

Frontispiece to Lyell's Principles of Geology

changed and seemed intermixed like Mary Anning's *Ichthyosaurus*. They did not fit in with Owen's concept of a 'separate creation' whereby God had created each of the animals and plants from scratch as a one-off. The Ichthyosaurs, like the later discovery of the bird-like *Archaeopteryx* in 1861, were important not just because they were extinct but because they seemed to represent transitional stages between species. Perhaps they became extinct because they could not compete with later, more efficient models.

The time was ripe

Darwin, the theological student, was fond of quoting from the scriptures on the *Beagle*. If he had quoted Ecclesiastes: 'To all things there is a season ... a time for every purpose and for every work', it would have been appropriate. The problems posed by discoveries like Mary Anning's and the geologists' provided the context for his ideas. It was as though the scientific world were preparing itself for Darwin.

Chapter 4

Barnacles and Rocks

The return of the *Beagle*

At last Charles came home. The *Beagle* docked on
2 October, 1836, nearly five years after it had set off.
On board were thousands of specimens, pickled,
dried or stuffed. The twenty-seven-year-old Darwin
had every reason to be full of himself – while still at
sea he learned that Adam Sedgwick had visited
Robert Darwin and told him his son would become
a leading scientist on his return. 'After reading
this letter', Darwin wrote, 'I clambered over the
mountains of Ascension with a bounding step, and
made the volcanic rocks resound under my geological
hammer'. He must have expected an enthusiastic
welcome home, especially in view of Sedgwick's visit.
Instead, as he wrote to a correspondent twenty years
later, his father had simply turned to Darwin's sisters
and observed quizzically: 'Why, the shape of his head
is quite altered.'

Hard work and recognition

Darwin was promptly elected to the Royal Geological
Society and indeed became its secretary; he
presented papers to the Royal Zoological Society and

was soon on the council of the Royal Geographical Society. By the time he was thirty he was a fellow of the august Royal Society itself. The two and a quarter years after he returned were the busiest of his life. He became a writer, and he liked it. Fitzroy had been engaged to provide an account of the *Beagle*'s voyage and of another earlier expedition. It was to

⊙ *Darwin in his late twenties by G. Richmond*

be a long book, and after Darwin's exertions it made sense for the Captain to ask him to write the third and final volume, the dull scientific record.

Darwin called his section 'Journal and Remarks', and it is still in print, though today it is published as *The Voyage of the Beagle*. He finished well ahead of Fitzroy, and had the frustration of waiting for his co-author to complete. Eventually in 1839 all three volumes came before the public. It was soon clear that Darwin's contribution most excited the scientific world and the general reader.

Working on the book had set Darwin thinking about the sheer diversity of the life forms, living and dead, that he had seen. But it was not just his own book that stirred his imagination.

Malthus and marriage

'In October 1838, that is fifteen months after I had begun my systematic inquiry, I happened to read for amusement *Malthus on Population*.'

Malthus is famous, indeed infamous, for this very book: his *Essay on the Principle of Population* (1798). In it he points out that nature produces vastly more offspring than can be allowed to survive. Regarding humans, Malthus states that poverty, early death and famine are the natural, inescapable, even God-given, result, since increase in food supply can never keep up with our over-production of babies. Darwin at once saw Malthus' point – which did not stop him from going on to sire ten children. In his autobiography, written for those children, Darwin said:

'Being well prepared to appreciate the struggle for existence which everywhere goes on, from my long-continued observation of the habits of animals and plants, it at once struck me that, under these circumstances, favourable variations would tend to be preserved, and unfavourable ones to be destroyed. The results of this would be the formation of a new species. Here, then I had at last got a theory by which to work.'

In 1839 Darwin married his cousin Emma Wedgwood, daughter of 'Uncle Jos', and would find great happiness in his marriage and family life with the eight survivors of their ten offspring; two, a boy and a girl, were to die young.

The geologist

Darwin's next three books were about geology. The first one covered coral reefs. Darwin was proud of having gained the friendship of his hero Charles Lyell. The great geologist was the leading expert on coral reefs, which he considered were formed round the rings of extinct underwater volcanoes. It made sense – coral couldn't grow in deep water, so something had to raise the seabed up to where there was enough light. Darwin had seen plenty of coral reefs, and he thought Lyell was wrong – that his theory didn't match the facts. Darwin realised that it wasn't a matter of raising the seabed but of lowering the sea level. In the distant past the sea had been vastly shallower, enabling coral to thrive. As it rose the coral had risen with it, raised not on volcanic craters but on the dead shells of its own ancestors. Actually, Darwin didn't get it quite right – he thought the sea level had risen because the seabed sank, but we now know that it was because of melting glaciers at the end of an ice age. But it fitted the facts better than Lyell's idea, and because Lyell was a true scientist he accepted Darwin's interpretation. Meanwhile, Darwin visited Glen Roy in Scotland and wrote a paper attributing its curious geological structure to the action of the sea, but realised he was wrong when another expert demonstrated it was caused by glaciers. Like Lyell, Darwin had the true scientist's objectivity.

Volcanoes, barnacles and the B notebook

The next two books were on the geology of real volcanic islands and of South America. He then became interested in barnacles. In 1844 his friend the botanist Joseph Hooker suggested he should study one species in depth. Charles took Hooker's advice, and chose barnacles. For eight years he studied every barnacle he could get his hands on, including barrels of Pacific barnacles supplied by the faithful Covington from Australia. Darwin wrote four monographs on the various types of barnacle, their sex life and their fossil ancestors. These highly specialised studies gave him considerable insight into how this one species had developed and the relations between living creatures and long dead fossils, and these conclusions helped to support a rather broader notion.

Close examination of barnacles suggested to him both how they were related to other shellfish and how they had come to diverge from these relatives. The notion chimed in with an idea that been ticking away at the back of his mind.

On his return from the *Beagle* voyage Darwin had written up two notebooks: the first, the 'Red Notebook' was the one he had kept to record his experiences on the *Beagle*, but he also started another one: small, pocket sized and leather-bound, the shorter 'B' notebook, written in scrappy, discontinuous phrases. On the first page he, curiously, wrote 'Zoonomia' the title of his grandfather's speculations on evolution. Inside he drew a diagram like a child's drawing of a spiky tree – he was by his own admission no artist – to illustrate how various species might have developed. Above it are the words 'I think'.

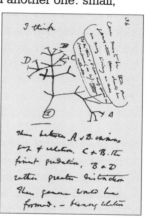

⊙ *Darwin's sketch, 1837, his first diagram of an evolutionary tree*

It was the first in a series of such notebooks in which he developed a revolutionary idea. In old age Darwin told his children how he started it in July 1837 when he had not yet been back six months:

'I opened my note-book for facts in relation to the Origin of Species about which I had long reflected, and never ceased working for the next twenty years.'

Chapter 5

The Origin of Species

A package in the post

Darwin did indeed work at his theory of natural selection for over twenty years. He wrote brief notes on it in 1838, an unpublished essay of thirty-five pages in 1842 and a 240-page unpublished essay in 1844. He was still struggling to reconcile his theory and Christianity in the 1844 essay:

'It accords with what we know of the laws impressed by the Creator on matter that the production and extinction of forms should, like the birth and death of individuals, be the result of secondary means. It is derogatory that the Creator of countless Universes should have made by His will the myriads of creeping parasites and worms, which since the earliest dawn of life have swarmed over the land and in the depths of the ocean.'

His grand plan was to publish a huge multi-volume tome at some unspecified future date. Then in June 1858 he got a package in the post. It was from Alfred Russel Wallace, a talented naturalist who, like Darwin, had worked in South America, where he collected specimens for resale to collectors, not having the advantage of a private income. Darwin had been impressed by his ideas and encouraged him. Wallace was now in the Malay Archipelago and while laid up with malaria had composed a twenty-two-page essay that he sent to Darwin for comment.

In Darwin's words, the essay 'contained exactly the same theory as mine'.

⌄ *Alfred Russel Wallace in 1862*

A joint presentation with neither present

Darwin was in despair. He complained to Lyell, 'All my originality, whatever it may amount to, will be smashed.' Nevertheless, he not only recommended to Wallace that he should publish his paper, but was even prepared to find an appropriate journal. Darwin's own altruism was matched by what he later called Wallace's 'generous and noble …

disposition'. Learning that Darwin had been working on the concept for twenty years, Wallace said that he had no wish to pre-empt him. In the event they agreed that Wallace's paper, together with one dashed off by Darwin, would be read at a special meeting of the Linnaean Society. Neither author was present: Wallace was abroad, and Darwin,

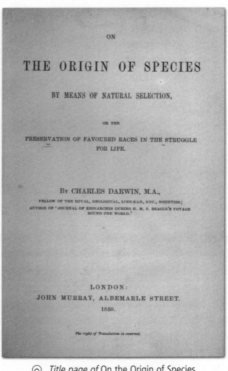

⊙ *Title page of* On the Origin of Species

unhappily, was attending the funeral of his baby son. Both papers were published in the Society's journal. They went virtually unnoticed except by a Professor Haughton of Dublin, whose verdict, in Darwin's words, 'was that all that was new in them was false, and what was true was old'.

The episode prompted Darwin to forget about the huge multi-volume masterwork and get on with what he called an 'abstract' of his ideas. He worked fast, and *On the Origin of Species* was published in the autumn of the following year, 1859. He sent a complimentary copy to Wallace with the note 'God knows what the public will think'.

On the Origin of Species: a chapter by chapter introduction

The introduction to *On the Origin of Species* harks back twenty years:

'When on board H.M.S. *Beagle*, as naturalist, I was much struck with certain facts in the distribution of the organic beings inhabiting South America, and in the geological relations of the present to the past inhabitants of that continent. These facts, as will be seen in the latter chapters of this volume, seemed to throw some light on the origin of species – that mystery of mysteries.'

He immediately goes on to mention Wallace and his 'excellent memoir' that had stirred him to put

his own ideas before the public. Nowhere does
Darwin concern himself with the beginning of life:
his subject is the development of species and the
interdependence of all nature – what we would
now call the 'ecosystem'. He gives the example of
mistletoe – how it needs the tree on which it lives
as a parasite, but also needs insects to pollinate
the flowers and birds to distribute its seeds. The
Introduction ends with a firm statement of his
position – species are capable of change, and a
key element in that change is natural selection.

> 'I am fully convinced that species are not
> immutable; but ... are lineal descendants of some
> other and generally extinct species. ...
> Furthermore, I am convinced that natural
> selection has been the most important, but not
> the exclusive, means of modification.'

Pigeon fancying

Chapter one is cunning: Darwin starts with
something everyone was familiar with and no one
could deny. Over centuries mankind has altered
animals to suit various human ends – horses have
been bred for strength or speed, cattle have been
bred for milk or meat, and dogs have been bred to
produce an amazing variety of shapes, sizes and
colours. But the animal he chooses to concentrate
on is the show pigeon. So-called 'pigeon fanciers'
ruthlessly cull the birds that do not have the
features they want and interbreed those with

features they do require. By this simple means they have been able to produce an astounding variety of colours and bizarre shapes. Darwin even joined two London pigeon clubs. He reports:

'The diversity of the breeds is something astonishing. Compare the English Carrier and the Short-Faced Tumbler, and see the wonderful difference in their beaks, entailing corresponding differences in their skulls. The Barb is allied to the Carrier, but, instead of a long beak, has a very short and broad one. The Pouter has a much-elongated body, wings, and legs; and its enormously developed crop, which it glories in inflating, may well excite astonishment and even laughter. ... The Jacobin has the feathers so much reversed along the back of the neck that they form a hood.'

And yet this whole outlandish miscellany is descended from the common rock pigeon. Darwin goes on to describe how breeders, not just of pigeons, but of horses and cattle produce their changes by an 'accumulation in one direction of differences absolutely unappreciable by an uneducated eye'.

Already he has demonstrated, from commonplace, everyday examples, how huge changes can be brought about in animals by the selection of tiny differences, which are then expanded over generations by continuing selection.

Incipient species

The whole concept of 'species', as propounded by
Linnaeus, was less than a hundred years old. The
Bible says nothing about species, just that God
created the different types of animals (what
Linnaeus categorised as 'genera') on different
days, and Adam gave them names. Darwin quotes
Wallace's work on butterflies to demonstrate that,
in practice, varieties, as in the hugely differing
varieties of pigeon or cattle, 'cannot be
distinguished from species'. The distinction is in
fact artificial, and many varieties are 'incipient
species'. Are Indian hump-backed cattle or Scots
hairy cattle the same species as Jerseys and
Herefords? And does it matter?

The struggle for existence

In the third chapter Darwin introduces 'the struggle
for existence', pointing out that because of this
struggle even small differences give certain
individuals the edge over others, thus ensuring their
survival. He introduces the term 'natural selection',
but says the philosopher Herbert Spencer's concept
of 'survival of the fittest' is equally valid. Referring
again to the pigeons, he says:

'We have seen that man by selection can certainly
produce great results, and can adapt organic
beings to his own uses, through the accumulation
of slight but useful variations... But Natural
Selection ... is a power incessantly ready for action,

and is as immeasurably superior to man's feeble
efforts, as the works of Nature are to those of Art.'

Techniques of survival vary. Many fish species
have survived by producing hundreds of eggs in
order that one or two may reach adulthood, while
penguins produce a single egg but protect it
carefully. And predators are essential to the balance
of nature. For example, the number of bees in a
district is related to the number of cats: the cats
keep down the field mice that would otherwise
destroy bees' nests and drive off the bees.

Infinitely varied diversities

In his Chapter 4 Darwin attributes the amazing
diversity of life forms to the 'infinitely complex and
close-fitting' mutual relation of 'all organic beings
with each other and their physical conditions of
life'. Each organism finds its own niche, a role to
which it is well suited, and is thus able to survive.
He shows how in social animals (like ants and
bees) natural selection adapts each individual for
the benefit of the whole community: the queen lays
eggs, the sterile workers look after the queen and
the larvae, the heavily armoured but sterile soldiers
defend the nest and the queen.

Sexual selection

But natural selection is not the only mechanism
of change. There is also sexual selection – the
'struggle between individuals of one sex, generally

the males, for the possession of the other sex',
which leads to the strongest males having more
offspring. Darwin was to elaborate this point in
more detail in *The Descent of Man*.

Isolation

Thinking back to the Galapagos Islands, Australia
and New Zealand, Darwin points out how isolation
can lead to the production of unusual and unique
organisms. As he remarks in a later chapter:

'Oceanic islands are sometimes deficient in animals of
certain whole classes, and their places are occupied
by other classes; thus in the Galapagos Islands
reptiles, and in New Zealand gigantic wingless birds,
take, or recently took, the place of mammals.'

Of course, once humans introduce new, more
efficient plants and animals, the native ones are
likely to suffer.

Difficulties

Chapter 5 covers adaptability, and in particular
Darwin gives examples of cases where organs that
have fallen into disuse tend to wither away – the
eyes of animals living in the dark, the wings of large
birds. This leads him into a remarkable series of
chapters in which he looks at potential difficulties
with his theory. He considers breeds of woodpecker
that don't peck wood, ducks and geese that don't go
near the water but still have webbed feet, and water

birds whose feet are not webbed. But he turns the
problem round and uses these examples to discredit
the idea of 'separate and innumerable acts of
creation'. What sensible creator would give webbed
feet to a land bird? But, seen as transitional stages
in natural selection, or even as features that
remained because they did not inconvenience these
birds in any way, they make sense.

With considerable insight, Darwin perceives the
importance of flexibility, observing that an organ
might take on a wholly new purpose; he suggests
that the swim bladder in fishes, which helps to keep
them afloat, became converted into the lungs used
by land animals to breathe. He considers the idea of
evolution through a series of sudden leaps forward,
as in Richard Owen's hypothesis that God introduced
distinct, fully developed new species every few
thousand years, and dismisses this because his own
studies of the selective breeding of pigeons, horses
and cattle had shown that evolutionary change is
minuscule, slow and subtle, and takes place
cumulatively over numerous generations. He looks at
hybrids and wonders why some crossbred plants and
animals, like the mule (offspring of a male donkey
and a female horse) are sterile, and considers
whether this undermines his theory.

Fossils and geology
Darwin had been deeply impressed by discovering
'great fossil animals' in South America, and recalled
how their resemblance to living creatures had

'haunted' him. But in *The Origin* he is concerned
that the study of fossils had not produced much
evidence of transitional forms, or halfway stages
between the extinct creatures and living animals.
He can only, rightly, attribute this to the paucity
of the fossil record. He is equally troubled by the
sudden appearance of a host of fossil forms at
certain times, notably during the 'Cambrian
explosion' (now put at about 540 million years ago).
Could this have been one of Owen's leaps forward?

Darwin's difficulty lay in his acceptance of Lyell's
dismissal of 'catastrophes'. We now know there have
been major catastrophes leading to the destruction
of a high percentage of life, and that afterwards the
surviving organisms flourished in an amazing variety
of forms, unrestrained by competition with the
supposedly more 'efficient' organisms that had been
wiped out. Without the destruction of the dinosaurs,
would mammals ever have come into their own?

All the time in the world

The importance of Lyell's work to Darwin was in its
revelation of the enormous antiquity of the Earth.
As Darwin says in his Chapter 11:

> 'He who can read Sir Charles Lyell's grand work on
> the Principles of Geology, which the future historian
> will recognise as having produced a revolution in
> natural science, and yet does not admit how vast
> have been the past periods of time, may at once
> close this volume.'

The whole key to natural selection is that it was able to work through tiny changes because it had plenty of time – eons of time, in fact all the time in the world. The Victorian guesses that Darwin quotes – such as four hundred million years since the formation of the Earth's crust – are gross underestimations, but the principle was sound.

Extinction and survival

Chapter 11 goes on to discuss 'the slow and successive appearance of new species' and points out that not all ancient species have died out – some shellfish, for instance, exactly resemble their fossil ancestors. They survive because they have not been superseded by more efficient species. Darwin then looks at the affinities between extinct species and living species, and remarks, in a sentence that virtually sums up his whole argument:

'The theory of natural selection is grounded on the belief that each new variety and ultimately each new species, is produced and maintained by having some advantage over those with which it comes into competition; and the consequent extinction of less-favoured forms almost inevitably follows.'

Geographical distribution

Darwin is troubled by how plants and animals got from one place to another across vast oceans. Or were similar species created at different points on the Earth's surface? He comes up with many

ingenious solutions, but is hamstrung by his
ignorance of 'Continental drift'. We now know
that at one time all the continents formed a single
mass, and that in various stages over the millennia
they slowly broke apart, carrying organic life
with them.

Divergence and similarity

In the fourteenth of his fifteen chapters, Darwin
returns to the enormous variety of organisms. He
elaborates how every species tries to find a niche
for itself; how because the varying descendants of
each species try 'to occupy as many and as different
places as possible in the economy of nature, they
constantly tend to diverge in character', and there is
a 'steady tendency in the forms which are
increasing in number and diverging in character, to
supplant and exterminate the preceding, less
divergent and less improved forms.' In short,
species become specialists by being flexible and
acquiring characteristics specific to their needs.
This leads to anomalies, such as whales looking like
fish, although they are actually mammals.

Many different insects are capable of
luminescence – that is, of creating light. The
obvious conclusion is that they share an ancestor,
but close study revealed to Darwin that, 'The end
gained is the same, but the means, though
appearing superficially to be the same, are
essentially different' – different parts of the body
light up; sometimes it is only the females that are

luminescent, sometimes only the larvae; sometimes the light even comes from luminous bacteria inside the abdomen. Here Darwin is describing what we now call 'convergent evolution'.

Then there is the phenomenon known as 'Batesian mimicry', after Henry Bates who first observed it while working with Wallace in South America. He noticed that several species of butterfly had evolved to look exactly like altogether different species – ones that were unpleasant-tasting to birds, and so were left alone by them. Butterflies not camouflaged in this way were eaten before they could reproduce, while the look-alikes went on to breed. Here was natural selection in action.

Finally in this chapter he takes on Richard Owen, who had written what Darwin calls 'an interesting work', *On the Nature of Limbs*. Owen had had difficulty in explaining why limbs doing completely different jobs should all be constructed in the same way – assuming each animal had been a bespoke creation by an intelligent Creator. Or as Darwin puts it:

'What can be more curious than that the hand of a man formed for grasping, that of a mole for digging, the leg of the horse, the paddle of the porpoise, and the wing of the bat, should all be constructed on the same pattern, and should include similar bones, in the same relative positions?'

What is more, he had found the same thing in Australia. Though the bone and musculature were different from those of European animals, the leaping kangaroo, the climbing koala and the ground-rooting bandicoot all had limbs constructed in the same way as each other.

What troubled Owen became perfectly simple once you accepted natural selection and inheritance from a common ancestor.

Conclusion

In his final chapter, Darwin describes the whole book as 'one long argument' and recapitulates what he has been saying. In essence these are the key points of his theory as it appears in *The Origin*:

- Darwin learned from Malthus that all organisms produce too many offspring. Unchecked population growth would quickly use up the Earth's natural resources.
- In practice, only a limited number of each species are able to survive long enough to breed. The rest die out. The same is also true of whole species, which can themselves become extinct.
- Organisms, even of the same species, have individual variations, for example in size and colour.
- These variations can to some extent be inherited.
- Humans use this natural variability to select the qualities they want and so to adapt plants and animals to their requirements.

- The inherited variations that survive in the wild are those that help adapt the organism or the species to its environment, thus providing a better chance of survival and reproduction.
- Over time, organisms that survive will be those that have become increasingly well adapted to their environment. The rest will die without reproducing themselves in sufficient numbers.
- Some individuals, among both plants and animals, are naturally stronger, more vigorous or more attractive in small ways – these have a sexual advantage, helping them to thrive and survive.
- Species are not fixed but can be altered. Just as human or natural selection can alter varieties of plants and animals so whole new species can be developed.
- Changes are small and gradual, but over enormous periods of time they can be immense.
- Nature is a complex, interrelated machine that works through the interaction of numerous plant and animal organisms.

As Darwin put it:

'If, then, animals and plants do vary, let it be ever so slightly or slowly, why should not variations or individual differences, which are in any way beneficial, be preserved and accumulated through natural selection, or the survival of the fittest? If man can by patience select variations useful to

him, why, under changing and complex
conditions of life, should not variations useful to
nature's living products often arise, and be
preserved or selected? What limit can be put to
this power, acting during long ages and rigidly
scrutinising the whole constitution, structure, and
habits of each creature, favouring the good and
rejecting the bad?'

Darwin was well aware that his ideas might
shock religious susceptibilities and was careful to
point out that:

'A celebrated author and divine has written to
me that, "he has gradually learned to see that it
is just as noble a conception of the Deity to
believe that He created a few original forms
capable of self-development into other and
needful forms, as to believe that He required a
fresh act of creation to supply the voids caused
by the action of His laws."'

The 'celebrated author and divine' was the
Reverend Charles Kingsley, author of *The Water
Babies*.

Others responded differently. Richard Owen
suddenly found he was getting a lot more visitors
to the Natural History branch of the British
Museum, many of them asking to see varieties of
pigeon. Karl Marx deplored the 'clumsy English

style of argument', but still felt 'Darwin's work is most important and suits my purpose in that it provides a basis in natural science for the historical class struggle'. The explorer and missionary David Livingstone declared that he had not been aware of any examples of 'survival of the fittest' in all the years he had spent in the African bush.

The most famous response of all was that of the biologist and anatomist Thomas Huxley. Huxley was an agnostic (he himself coined the word), and upon reading *The Origin of Species* he is said to have simply remarked, 'How extremely stupid not to have thought of that!' Huxley was to play a major role in defending Darwin; but before looking at the debate instigated by the publication of *The Origin* important developments on the continent of Europe need to be considered in the light of the hypothesis it put forward.

Chapter 6

Lamarck and Mendel

Lamarck

The publication of *On the Origin of Species* established Darwin's concept of natural selection in the public mind, together with Herbert Spencer's phrase, 'survival of the fittest'. But how did a particular specimen become the 'the fittest'? What were the mechanics of change? Darwin knew the answer proposed by 'The Chevalier Jean-Baptiste Lamarck'. Lamarck lived from 1744 to 1829, dying when Charles was already at Cambridge. He was a French aristocrat, but so poor that he failed to attract the notice of the Jacobins and survived the Revolution with his head on his shoulders.

⊙ *A late-19th-century engraving of Lamarck*

Lamarck believed in the transmission of inherited characteristics. In his own words:

'All which has been acquired, laid down, or changed in the organisation of individuals in the course of their life is conserved by generation and transmitted to the new individuals which proceed from those which have undergone those changes.'

To take a simple example: a blacksmith works hard and gets strong arms; his son will inherit this strength; the son works hard as a blacksmith and gets even stronger arms, which he passes on to his own son. Animals, too, change through their own efforts and pass on the changed characteristics to their offspring. To be fair, Lamarck did not suggest animals 'willed' themselves to improve, as is often suggested. He used the word *besoin*, meaning 'need'. Wading birds needed to stretch to keep their feathers dry, and stretched generation after generation so their legs got longer. The lizards that became snakes found having legs inhibited them when slithering through the undergrowth, so they needed to reduce their legs generation after generation until they eventually disappeared. Over a long period of time the changes could be drastic – a four-legged mammal turned itself into the fish-like whale because it *needed* to adapt to living in the sea.

A justly celebrated naturalist

His critics often claim that Lamarck said giraffes got their long necks by stretching up to nibble the high branches. He did not; Lyell made this suggestion ironically in a prolonged attack on Lamarck. Darwin himself half accepted the 'inheritance of acquired traits'. It makes a lot of sense, but it is totally untrue: I recall at Cambridge a genetics lecturer pointing out that for thousands of years he and all other Jewish males had been circumcised. According to Lamarck, this operation would eventually be unnecessary. Yet Jewish men are still born with foreskins; or, as he told his students, quoting the words of Hamlet, 'There's a divinity that shapes our ends, Rough-hew them how we will ...'

Darwin was far less hostile to Lamarck than Lyell was, and recognised his importance in developing the concept of evolution. In 1861 Darwin wrote:

'Lamarck was the first man whose conclusions on the subject excited much attention. This justly celebrated naturalist first did the eminent service of arousing attention to the probability of all changes in the organic, as well as in the inorganic world, being the result of law, and not of miraculous interposition.'

In other words, Lamarck prepared the way for *On the Origin of Species*.

Pangenesis

Neither Lamarck nor Darwin, however, had any proper evidence of the mechanism by which evolution worked. Darwin's explanation was just as wrong as Lamarck's, and it wasn't even original.

The best Darwin could come up with was a hypothesis that had been around for two thousand years: Hippocrates' so-called 'Pangenesis'. According to this theory, every part of the body exudes a little bit of itself. They all come together in some liquid (the Greeks thought it was blood, but Darwin's cousin Francis was to disprove this) that transfers them to the offspring. Some people even claimed to have seen tiny human beings (*homunculi*) in sperm examined under a microscope.

Darwin's pigeon breeding might have given him a clue to a better hypothesis. In Chapter 1 of *The Origin* he tells how he crossed 'a Spot – a white bird with a red spot on the forehead' with 'mongrels' that were 'dusky and mottled', re-crossed their offspring, and suddenly got a bird 'of a beautiful blue colour', although there had been no blue in its ancestors for generations back. He simple calls this 'reversion to ancestral characters'. Had he worked his way meticulously back through the generations of pigeons he might have got a clue to the mechanism by which natural selection works.

Alternatively he could have taken one the new steam ferries to Calais and then, travelling by

Moravia, in the Austrian Empire. Here he would have found a monk doing the work for him, but with peas rather than pigeons.

'Insight and clarity of knowledge'

In 1822, when Darwin was at public school in Shrewsbury, Gregor Mendel was born in Silesia to a family of poor farmers. He had a burning desire to learn, and as his parents could not afford to send him to university he joined the Augustine Abbey of St Thomas in Brünn (today, Brno) in the hope of getting the Church to pay for his education. His prayers were answered, and in due course the abbot sent him off to Vienna University to learn to be a teacher. While he was there he became interested in science, but failed his exams. His examiner commented, 'He lacks insight and the requisite clarity of knowledge'. So in 1853 Mendel returned to the monastery. He was thirty-one; Darwin at that age had already published *The Voyage of the Beagle*. The Augustinians were a teaching order, and Mendel had failed his teaching examinations. What was he to do?

Peas

Mendel was set to work in the monastery garden. He decided to use the opportunity to pursue his interest in science, and in particular to study how variations arise in plants. After considering several species he decided to grow peas – ordinary, edible peas. Between 1856 and 1863 he grew 29,000 pea plants, and created a controlled and limited experiment, of which he kept meticulous and detailed records that enabled him to draw conclusions.

He decided on peas because certain qualities of pea plants came in pairs:

1. The flower was either red or white.
2. The flower was either in the middle of the stem or at the end of it.
3. The stem was either long or short.
4. The seed shape was either round or wrinkled.
5. The seed was either yellow or green.
6. The pod shape was either swollen or constricted.
7. The pod colour was either yellow or green.

He called each of these alternatives a 'factor'. A pea plant had a red flower factor or a white flower factor, a tall factor or a short factor, and so on.

He crossbred the peas by the normal method used by gardeners – carefully transferring pollen containing male cells from the stamen (the 'male' reproductive organ) of one pea plant to the pistil (the 'female' reproductive organ) of a second plant. He called the male and female elements 'gametes', from the Greek for 'husband/wife'.

Dominance and recession

Now if a red-flowering plant is crossbred with a
white one, you might expect the new flower to be
pink. Darwin certainly would have done. But no,
the 'offspring' were all bright red. So it looked as
though the red had wiped out the white.

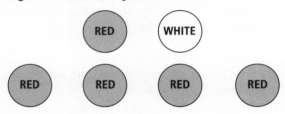

Mendel then crossbred these red flowers and
suddenly got the occasional white – just as
Darwin had been surprised to find a blue pigeon
hatching when neither of its parents was blue.

So, what was happening? Mendel grew a
great many peas before he worked it out.

The red factor was 'dominant', but the white
factor wasn't destroyed. It remained dormant, as
part of the make-up of the plant, but it didn't show
itself. Mendel called it 'recessive'. The offspring
may have looked like purebred red flowers, but
actually they were hybrids, containing 'factors'
derived from each of their pure red and white
parents. Then he crossbred two of these red
coloured hybrids – A and B. Here they are, each
with two plant colour factors:

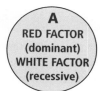

Mendel found that the four different factors came together in four different ways and produced four types of offspring:

C, D and E are all red because they contain the dominant red factor, but F has no dominant red so it reverts to being a purebred white.

Now, in each pair of qualities there will be a dominant and a recessive factor: for example tall dominates short, and in the seeds yellow dominates green.

Peas produce two crops a year, so the possible variations became enormous, but, as Mendel found, the variations were always in the ratio of three dominant to one recessive. The important point for a scientist was that it was predictive. Mendel could actually predict what his results would be. Darwin was never able to do that with his law of natural selection, and that was to prove a major weakness.

The Father of Genetics

Mendel had discovered what we now call 'genetics': the mechanism that Erasmus Darwin, Charles Darwin, Charles's cousin Francis, and Lamarck all sought in vain. And he gave his solution to the world. Perhaps not quite to the world: just to Moravia, or, more precisely, to Brünn.

He published his *Experiments with Plant Hybrids* with all his calculations in the *Proceedings of the Natural History Society of Brünn*. It was 1866, Darwin had published *On the Origin of Species* seven years before and Mendel had read it and admired it, though he did not always agree with what it said, especially about Pangenesis.

⊙ *Gregor Mendel*

Mendel's Laws

Three scientific laws were later to be deduced from Mendel's findings, the first two based on his propositions and the third based on his findings:

1. **Mendel's Law of Segregation** says that the two factors in each plant always split up, so creating the four combinations shown above.
2. **The Law of Independent Assortment** says that just because a plant has, say, two dominant red factors, this has no effect on anything else, such as seed colour or height. This means that there

can be a large variety of different offspring: tall red with yellow seeds; tall white with yellow seeds; short red with yellow seeds; short white with green seeds, and so on.

3. **The Law of Dominance**; one factor is always dominant.

Lost without trace

Darwin was an honest man. Having admitted in *The Origin* that there were unsolved problems, he suggested that future scientists might find the solutions. Had he subscribed to the *Proceedings of the Natural Science Society of Brünn,* he would have discovered that his biggest question had already been answered. Alas, though Mendel had heard of Darwin, Darwin never heard of Mendel. Over the next thirty-four years Mendel's brilliant thesis would be quoted only three times in scientific papers. When he did come into his own and his genius was recognised, he had already been dead sixteen years. Mendel did not consider his life a failure, though. In 1868 the failed undergraduate was elected Abbot of his monastery. He threw himself into the leadership of the monastery with the same zeal as he had cultivated and recorded his peas, and became an excellent abbot.

Mendel's story demonstrates how important rapid, unrestricted, worldwide communications have become to science. Today Mendel would have published his findings on the Web, and within hours the news would have reached Darwin.

Chapter 7

Apes and Humans

Precursors

On the Origin of Species had a print run of 1,500. It sold out at once; copies were even on sale to commuters at Waterloo Station, who snapped them up, and John Murray, the publisher, immediately ordered a further print-run of 3,000. This sounds impressive until it is set against the rapid sales of 21,000 copies of a book called *Vestiges of the Natural History of Creation*, published anonymously in 1844. *Vestiges*, which everyone knew was by the Scottish journalist Robert Chambers, anticipated some of Darwin's ideas, but Chambers also promulgated humankind's connection with the apes, although without any proper scientific basis.

Evolution and apes went together in the public mind, and no

⊙ *Caricature in* The Hornet Magazine, *1871*

sooner was *On The Origin of Species* printed than critical reviews were taking it for granted that the book was about how men are derived from monkeys. Punch and other humorous periodicals delighted in cartoons portraying Darwin as an ape. Yet Darwin never suggests any link between humans and apes in *The Origin*. While it may well have been implicit in his argument, this association of his ideas with the man/ape debate was largely due to the popularity of *Vestiges* and to his friend Thomas Huxley's involvement in the controversy.

Owen's denial

Richard Owen, who saw himself as the leading scientific figure of the age – he even taught Queen Victoria's children biology – took up the task of refuting this dangerous heresy. In 1854, he gave a talk to the British Association in which he denied that apes, such as the newly discovered gorilla, could stand on two feet or 'transmute' into humans. Owen claimed that the human brain has important unique structures absent in ape brains.

Darwin had condemned the pseudo-science of *Vestiges*. Nevertheless, at the time of Owen's

⊙ Sir Richard Owen in old age

lecture he wrote to Thomas Huxley, 'I cannot swallow Man [is] as distinct from a Chimpanzee as an Ape from a Platypus'. In March 1858 Huxley himself counter-attacked Owen in a lecture in which he pugnaciously asserted that the 'mental & moral faculties are essentially ... of the same kind in animals and ourselves'.

'Darwin's Bulldog' versus the 'mendacious humbug'

It was in the following year that *The Origin* was published, and since it was known that Darwin and Huxley were close associates, it was assumed, rightly as it happens, that Darwin shared his views on mankind's ancestry. In fact Huxley had reservations about some aspects of natural selection; nevertheless he became Darwin's most zealous defender, so much so, that he referred to himself as 'Darwin's Bulldog'. To Huxley's delight he was asked to review *The Origin* in the *Times*, and came down thunderously in praise of it.

Not all reviews were so favourable. Richard Owen had been strongly supportive of Darwin when he had first returned on the *Beagle*, and had even at first welcomed *The Origin*, believing it to be an extension of his own ideas. When Darwin disabused him, Owen went over to the opposition. Owen wrote long article in the *Edinburgh Review*, then an extremely influential quarterly. He began by praising Darwin for his work on barnacles and geology, and this is significant because, unlike Chambers, Darwin was an important and well-known scientific figure – his views had to be taken seriously by the scientific community. But Owen then goes on to describe the new book as an 'abuse of science'. The article, as was the custom, was published anonymously. This gave the writer the opportunity to exalt the work of a certain Professor Richard Owen as a much more scientific explanation, and one that still left plenty of room for God. Darwin was very hurt, and told Huxley, 'It is painful to be hated in the intense degree with which Owen hates me.' Huxley, on the other hand, loved a fight. Owen called Huxley an 'advocate of man's origins from a transmuted ape', while Huxley gleefully told Darwin, 'Before I have done with that mendacious humbug I will nail him out, like a kite to a barn door, an example to all evil doers.

⊙ *Thomas Huxley in 1874*

'Darwin's Bulldog' versus 'Soapy Sam': the Oxford debate

In 1860 a splendid and famous debate took place in Oxford at the Museum of Natural History. Darwin avoided this sort of thing so, as usual, Huxley took up the cudgels on his behalf. His opponent was the

Bishop of Oxford, Samuel Wilberforce, universally known as 'Soapy Sam', who thirteen years before had led the attack on *Vestiges* with considerable success. The debate is chiefly remembered because Wilberforce supposedly asked Huxley whether it was through his grandfather or his grandmother that he claimed descent from a monkey. Huxley thought he replied something along the lines of 'he would not be ashamed to have a monkey for his ancestor, but he would be ashamed to be connected with a man who used his great gifts to obscure the truth', at which Lady Brewster is said to have fainted. But since everybody was in a high state of excitement and nobody took any notes, no one is

⊙ *Caricature of Samuel Wilberforce in* Vanity Fair, *1869*

sure what really happened. Joseph Hooker, who was in the audience to support his friend Darwin, was firmly convinced that it was his own intervention that won the day for Darwin, while Wilberforce was certain he himself was the victor. But, as one observer remarked, 'everybody enjoyed himself immensely and all went cheerfully off to dinner together afterwards'.

Fitzroy again

In fact not quite all. Robert Fitzroy was there, now an Admiral. He had come to give a talk on storms –

he was virtually the founder of the new science of meteorology – but he chose to join in the evolution debate. Hooker describes how 'a grey haired Roman nosed elderly gentleman … lifting an immense Bible first with both and afterwards with one hand over his head, solemnly implored the audience to believe God rather than man. *The Origin of Species*, Fitzroy said, had given him 'acutest pain'. The audience roared him down.

⊙ *Lithograph of Robert Fitzroy, c. 1845*

It was all too much for Fitzroy. He sank into depression, and five years later killed himself. It is a

cruel irony that, as a suicide, the Christian
fundamentalist had to be buried in unhallowed
ground but the agnostic, probable atheist Darwin,
who wanted no more than to join the worms quietly
in his garden at Down, was instead when the time
came given a full state funeral at Westminster Abbey.

Neanderthal Man

In 1856 the skeleton of an early form of man was
found in the Neander Valley near

Düsseldorf in Germany. In 1863 it
was named Neanderthal Man.
Fundamentalist commentators
saw it as a human, albeit one
severely distorted by congenital
disease. Others claimed it was
our apelike ancestor. Huxley,
with Darwin's approval, argued
that it occupied a lower
position than *Homo sapiens*.

⊙ *First reconstruction of
Neanderthal Man*

Not an ancestor, but a cousin
that had become extinct. True humans, they said,
had originated in Africa.

'The Bulldog' triumphant

Huxley proved to be an abler anatomist than
Richard Owen. In 1862, working with the brilliant
surgeon and zoologist William Flower, he gave a
public dissection of an ape's brain. They showed

that it did have the very structures Owen said were not there. And so the running battle went on for years, with Huxley generally getting the better of the argument.

Owen's star began to decline. He had done good work in founding the natural history collection of the British Museum and in January 1863 he bought the *Archaeopteryx* fossil for the museum. It wonderfully fulfilled Darwin's prediction, that a proto-bird with unfused wing fingers would be found. Owen dogmatically insisted it was a proper bird, putting his convictions before his observation as a scientist. In 1871 Owen was found to be involved in a plot to end government funding of Hooker's botanical collection at Kew, and to turn the site instead into a public pleasure garden and/or a sewage farm, in order to promote his own plans for a national Natural History Museum. Additionally, he had always been notorious for stealing other people's ideas, and was dismissed from the Councils of both the Royal Society and the Zoological Society for plagiarism.

But at last, at Easter 1881, Owen's great monument, the magnificent, cathedral-like Natural History Museum, opened in South Kensington. It housed the huge fossils that Owen had named 'dinosaurs', and his statue in fine black marble for years dominated the head of the main stairs. In 2009 it was replaced by the white marble statue of Charles Darwin. Huxley's bones must be rattling in triumph in their godless grave.

Chapter 8

The Descent of Man

Shortly after the publication of *The Origin* Darwin wrote to Professor Baden Powell, professor of mathematics at Oxford and a noted liberal thinker:

'In the *Origin of Species* the derivation of any particular species is never discussed, yet I thought it best, in order that no honourable man should accuse me of concealing my views, to add that by the work "light would be thrown on the origin of man and his history". It would have been useless and injurious to the success of the book to have paraded, without giving any evidence, my conviction with respect to his origin.

However, ten years of Huxley and Owen sniping at each other made Darwin change his mind, and 1871 saw the publication of Darwin's most controversially titled work, *The Descent of Man, and Selection in Relation to Sex.*

It took him three years, even though this time he was working with no new evidence, but simply re-evaluating what was already known. Right at the start Darwin takes the bull by the horns and gives his readers the following rationale:

'... [the] sole object of this work is to consider, firstly, whether man, like every other species, is descended from some pre-existing form; secondly, the manner of his development; and thirdly, the value of the differences between the so-called races of man.'

A notorious similarity

He examines the similarities between what he calls 'the lower animals' and man including the 'notorious similarity' between human skeletons and those of other animals. In particular he looks at how akin we are to apes in our skeleton, our physique, our diseases and even, among many other things, in our

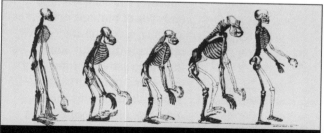

Frontispiece to Huxley's Evidence as to Man's Place in Nature, *1863*

liking for 'tea'. He notes that man retains vestiges of features found among the apes – he calls them 'abnormal reversions': wisdom teeth and residual body hair, the ability some of us have to move our scalp and ears, as well as features that seem to link us to animal instincts, like the often observed connection between the sense of smell and memory.

Gills and a tail

What especially interests him is the similarity of the human embryo within the womb to those of other

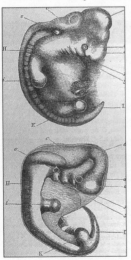

⊙ *Woodcut depicting the similar appearance of a human embryo (top) and a dog (bottom)*

mammals, like dogs, and the way, for the first few months of their existence, all embryos go through the same sequence of development. It was well known that at certain points in its growth the human embryo has what appear to be rudimentary gills and a tail. Darwin points out that the embryos of humans and animals, although they grow into vastly different adults, are for a long period during their development remarkably similar, and he sees this as evidence of their preserving the structure of our common ancestor, or 'progenitor'.

Animal intelligence

Darwin concludes that, 'Man is no more ... than a more highly organized form or modification of a pre-existent mammal'. He even goes so far as to guess at the nature of this creature: a ' hairy, tailed quadruped, probably arboreal in its habits, and an inhabitant of the Old World.' He insists, in fact, that there is a 'community of descent' among all vertebrates.

He next considers the apparent intellectual superiority of humans. This had been proposed as a way of showing that humankind is completely distinct from members of the animal kingdom that can only act on instinct. Darwin suggests that apes in the wild do not merely react instinctively – they are, for instance, capable of learning which plants are poisonous. He refers to Wallace with regard to the idea of man as a toolmaker, and discusses Dean Paley's example of birds migrating and building nests. He will later write about how animals can express feelings physically, but for the moment only remarks how animals can clearly feel pleasure and pain, can be happy and unhappy, suspicious or brave.

This leads him to remark that, though the

⊙ *Illustration from Darwin's* Descent of Man

mental powers of the higher animals are much less than ours, they are related to ours and are similarly capable of learning and development. He points out that the difference in mental powers between humans and apes is much less than that between apes and fish. More significant is the assertion that animals' mental powers are capable of advancement. He insists that the higher animals, like humans, go through a process of experiencing things, examining this experience and 'forming new resolutions as a consequence'. But some individual animals are cleverer than others, and this helps them to survive. The same is true of mankind.

When it comes to language his attitude is surprisingly Lamarckian, in that he seems to accept the inheritance of acquired characteristics, stating that the human brain is so large partly because 'the continued use of language will have reacted on the brain and produced an inherited effect'.

Soldiers of the Queen

Darwin sees man's moral ability to put himself at risk for the sake of others as related to the way social animals are also impelled by an instinct to aid the members of their community. Human history is full of noble examples of heroes laying down their lives for their country, their family, their regiment or their beliefs. Human martyrdom is nothing, however, to that of the soldier ants and termites that readily sacrifice themselves in their thousands for the sake of their anthill and their queen. They are not even doing this for their own posterity, since the soldiers are neuter – so why do they do it? In this instance, natural selection was assumed to act for the good of the species or of the group. As a trained theologian, Darwin was naturally concerned about morality and the human conscience, and plays it safe by commenting that our social instincts are much more highly developed than those of the kamikaze termites:

'So far as the highest part of man's nature is concerned there are other agencies more important. For the moral qualities are advanced, either directly or indirectly, much more through the effects of habit, the reasoning powers, instruction, religion, etc., than through natural selection ...'

But then he gives himself away by adding:

'Though to this latter agency [survival of the species] may be safely attributed the social instincts which afforded the basis for the development of the moral sense'.

The 'selfish gene'

Recent research has established that 'survival of the fittest' acts on individuals, and cannot be shown to act on any larger entity, such as the species or the anthill. So why do Guards officers, saints and termites sacrifice themselves? The best answer to the dilemma of self-sacrifice so far proposed is that put forward by Richard Dawkins in *The Selfish Gene*. The soldier termites share their genes with the queen, and it is the survival of these genes that matters. Instead of the gene being the means by which the termites reproduce themselves, the termites are the 'vehicle' that enables the genes to replicate themselves.

In the case of the Guards officer and the saint, Dawkins also postulates a kind of shared cultural inheritance, which he calls a 'meme', but this hypothesis is outside the remit of this brief book.

The God within us

In a challenging section towards the end of *The Descent of Man*, Darwin states, 'I am aware that the assumed instinctive belief in God has been used by many persons as an argument for His existence ...' He then proceeds to assault this view with the words:

'But this is a rash argument, as we should thus be compelled to believe in the existence of many cruel and malignant spirits, only a little more powerful than man; for the belief in them is far more general than in a beneficent Deity. The idea of a universal and beneficent Creator does not seem to arise in the mind of man until he has been elevated by long-continued culture.'

This is something very like Dawkins's 'meme'.

Survival of the fittest

Malthus's theory of population had been the trigger that set Darwin thinking about natural selection, and it reappears in *The Descent of Man*. Because man tends to breed more children than there is food available for, he too is subjected to a struggle for existence, with the result that natural selection comes into play:

> 'We can see this in the rudest state of society: the individuals who were the most sagacious, who invented and used the best weapons and traps, and who were best able to defend themselves, would rear the greatest number of offspring. The tribes which included the greatest number of men thus endowed would increase in number and supplant other tribes.'

Sexual selection

However, much of the book is taken up not so much with natural selection as with sexual selection. Darwin differentiates between the two by stating that, while natural selection depends on the successful survival of both sexes in a hostile world, sexual selection depends on the triumph of certain individuals over others of the same sex in relation to the propagation of offspring. He is well aware that lower animals like his beloved barnacles, which are rooted to one spot, have little opportunity for sexual

Male Brenthidae *beetles fighting over a female*

selection, especially if, like many barnacles, they have both sexes combined in the same individual. It is clear, however, that among the higher animals in the wild, such as lions or red deer, it is the strongest, dominant male that breeds with a variety of females while the weaker males remain unmated. In fact we now know that one reason for giraffes having long necks is that long-necked male giraffes are dominant.

But it isn't all about simple bullying; beauty matters too. In other sections of the book Darwin goes into detail about the relevance of sexual display among colourful birds and baboons with blue behinds.

Drawing of a peacock feather from The Descent of Man

A major advantage of sexual selection is that it enables two unrelated animals to reproduce, thus reducing the danger of hereditary disease. Darwin married his cousin, and two of his children died very young. Eight out of ten was actually a good survival rate in Victorian times, but Darwin was worried by the possible effects of inbreeding, as is evidenced by a rather bitter aside:

'When the principles of breeding and inheritance are better understood, we shall not hear ignorant members of our legislature rejecting with scorn a plan for ascertaining whether or not consanguineous marriages are injurious to man'.

Race

The third question that Darwin said he would address in *The Descent of Man* was 'the value of the differences between the so-called races of man'. Just as the Victorians were conditioned to feel horror at any hint of atheism, so we bristle at any hint of racism, but it was an accepted concept at the time, and it is interesting that Darwin refers to 'so-called races'. Both sides of Darwin's family were strongly opposed to slavery, indeed his son William once remarked that nothing moved his father more than cruelty to animals and slavery, as is demonstrated by his set-to with Fitzroy aboard the *Beagle*. Before that, in Edinburgh, one of his close friends had been the freed Guyanese slave who taught him how to stuff birds and other animals.

Believers in a separate creation had often claimed that the various 'races' were distinct species: Cuvier identified three such species, Kant four and Desmoulins twenty-two. In *The Descent of*

Man Darwin attacks these distinctions and insists that we all sprang from a common ancestor. He carries his argument further by saying that we humans owe it our species to recognize all other people as members of the same society and support them, although he is more than aware that the history of mankind reveals a very different picture.

The ascent of man

Nevertheless, Darwin was a man of his time, and in applying natural selection to human society he accepted the superiority of Western Europeans, if not over other races, at least over primitive society. He looks at various qualities, discusses their value in terms of natural selection and suggests that we see 'savages' as immoral because their social instincts are restricted to their own tribe, and their powers of reasoning are limited. He tells of his amazement, after his sheltered life in England, when he first encountered the people of Tierra del Fuego, in South America:

> 'These men were absolutely naked and bedaubed with paint, their long hair was tangled, their mouths frothed with excitement, and their expression was wild, startled, and distrustful.'

Interestingly, he says that instead of thinking how different they were, the first thought that came to his mind was 'such were our ancestors'.

He ascribes the extinction of ancient peoples to competition and the struggle for existence and ponders on why contact with 'civilised' nations can lead to the extinction of primitive societies. In the section 'Natural Selection as affecting Civilized Nations' he looks at changes effected in the process of natural selection by civil society. 'Civilization' has made it possible for the physically weaker members of society to survive and multiply while the cult of war kills off a proportion of the strongest. Darwin is concerned about the effect of this on natural selection, though he is not willing to accept drastic measures to remedy it.

Finally, he concludes:

'Man may be excused for feeling some pride at having risen, though not through his own exertions, to the very summit of the organic scale … We must, however, acknowledge, as it seems to me, that man with all his noble qualities, with sympathy which feels for the most debased, with benevolence which extends not only to other men but to the humblest living creature, with his god-like intellect which has penetrated into the movements and constitution of the solar system – with all these exalted powers – Man still bears in his bodily frame the indelible stamp of his lowly origin.'

Chapter 9

Disciples and Rivals

Darwin was to write many more books on topics ranging from *Movement in Plants* to *The Expression of Emotions in Man and Animals* and *The Formation of Vegetable Mould through the Action of Worms* before he died in 1882.

In 1859 the Prime Minister, Lord Palmerston, had suggested to Queen Victoria that Darwin deserved a knighthood, but the bishops, led by Wilberforce, opposed it. Paradoxically, by 1882 the religious establishment had become sufficiently reconciled to its errant son to agree that, despite Darwin's explicit request to be interred at Down House, he should be buried in Westminster Abbey. There Darwin lies still, next to Sir Isaac Newton, while both men's ideas have gone bowling on, being revised, supplemented and adjusted, but never discredited, and central to our understanding of the universe in which we live.

Embryos and the Missing Link

One of Darwin's most remarkable followers was the Prussian Ernst Haeckel (1834–1919), who followed up Darwin's thoughts about the similarity of the embryos of different animals. His 'recapitulation theory' suggested that the development of the embryo mirrors the evolution of animal life. It starts as a simple structure of a few cells, develops into a more complex group of cells, then into a curled, eyeless, boneless creature, which grows a backbone, a tail and so on, until at a comparatively late stage each embryo takes on the characteristics of its own species.

It was a highly ingenious and somewhat romantic concept, for which there was little hard evidence other than the obvious superficial similarities, and he was accused of fraudulently exaggerating these. Yet, while Haeckel over-simplified, his work has to some extent been justified by modern research into genes. All life carries with it a DNA record of its origins, and there is a high degree of overlap between very different creatures and even separate life forms, especially in their early stages. Also, as he proposed, if some part of the body has disappeared in the course of evolution it can often reappear and then later disappear again during the development of the embryo. An obvious example is the whale: whales don't have legs but, having evolved from land mammals, they do have tiny residual leg bones

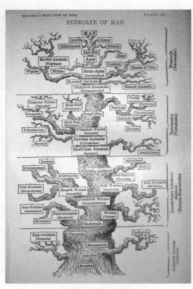

⌄ *Haeckel's Tree of Life*

deep in their bodies. During embryonic development, legs of sorts grow, then recede; and like all mammal embryos whale embryos also have hair at one stage, but lose it later.

Haeckel was a talented artist and used his artistic skill to illustrate his theories. He elaborated the idea of the Tree of Life that had appeared in scrappy form in Darwin's B notebook, but which Haeckel developed into the elegant forerunner of the trees of life that today appear in so many textbooks.

Ape-men

Leaving aside the Neanderthal skeleton, Haeckel reasoned that there must have been a creature halfway between a human and an ape – a 'missing link' – so he painted what he believed such a creature would look like, and named it

Pithecanthropus, or 'ape-man'. Without any real evidence, he suggested it would be found in the East Indies. Lo and behold, in 1891 one his students, in Java, found the top of a skull, a few teeth and a femur of *Pithecanthropus*, though we now call it *Homo erectus*: upright man. Together with the later discovery of 'Peking Man' the find suggested that mankind had originated in Asia. More recent discoveries in Africa, however, strongly suggest that *Pithecanthropus* was a blind alley – that the breed died out and we are actually descended from Africans, as Darwin had postulated.

Weismann and Mice

A very different follower of Darwin was the German biologist August Weismann, notorious for apparently taking the nursery rhyme 'Three Blind Mice' too seriously and cutting the tails off thousands of mice in order to test Lamarck's idea of the inheritance of acquired characteristics. If Lamarck was right the mice, after numerous generations, ought to have been born without tails, but this never happened.

⊙ *August Weismann*

Soma and germ-plasm

Weismann rejected Darwin's Pangenesis – the idea that each part of the parents' bodies contributed to the embryonic offspring. Writing a year after Darwin's death, he postulated what he called the 'germ-plasm' theory.

Weismann's experiments suggested that:

- Some cells are body cells; he called them, simply, soma (*soma* is Greek for 'body').
- Other cells are sex cells. He called the sex cells germ-plasm ('plasm' comes from the Greek *plassein*, to mould, and 'germ' from the Latin *germen*, sprout, bud).
- The soma (body cells) have nothing to do with the sex cells (the sperm and ova), and cannot affect them in any way. There is a 'barrier' between the two types of cell.
- The body can change: exercise can enlarge our muscles; mice can have their tails cut off; but no characteristics acquired by the body can be passed on to the wholly separate 'germ-plasm'.
- The germ-plasm contains the basic hereditary material – the information that ensures we give birth to humans and mice give birth to mice.

Otherwise Weismann was a devoted Darwinist and saw natural selection as the only possible force of evolutionary change. But, like Darwin, Weismann was a true scientist: he made numerous experiments to try to find an example of the inheritance of acquired characteristics that would contradict his own ideas. Hence the thousands of de-tailed mice.

Butterflies

Weismann also carried out important research on Batesian mimicry, studying Mediterranean butterflies that had evolved to resemble breeds that were unpleasant-tasting, and so were left alone by birds. The mimics survived, and were able to reproduce. As Darwin had suggested, they demonstrated natural selection.

Mendel recognised

Then Mendel's 1866 paper in the *Proceedings of the Natural History Society of Brünn* suddenly surfaced.

In 1900 Mendel's theory was 'rediscovered' by three scientists, the most significant of whom was Hugo de Vries, who took a leading role in applying Mendel's discoveries. However there was another key figure who was none of these three. This was an Englishman,

⊙ *Hugo de Vries*

William Bateson, Master of St John's College, Cambridge. He coined the words 'genetics' and 'gene'. Genes are Mendel's 'factors', the means by which characteristics are passed from one generation to the next. Genetics is the study of how reproduction and evolution works.

One might think that Mendel's evidence would have been seen as justifying Darwin's theories – the scientific proof he never found. In fact it was held to justify rival theories propounded by Bateson and De Vries.

Leaps and changes

In 1894, six years before Mendel's work reappeared, Bateson had published *Materials for the Study of Variation: treated with special regard to discontinuity in the origin of species*. It was a book of monstrostities, cataloguing monstrous features in animals such as humans with extra nipples or six toes, or bees with legs where their antennae should have been. Bateson said these 'variations' (Darwin had used the same word) were the real cause of evolutionary change. Darwin had considered this idea but rejected it.

Then in 1900 Bateson learned from De Vries and the others about Mendel's research. Bateson studied Mendel's writings and decided that his theories provided the structure he need to prove he was right.

Darwin's concept of natural selection was of a gradual process of small variations, but Bateman believed his monstrous variations worked quickly, and could explain evolution without any need for natural selection. His theory was called 'saltationism' (from the Latin *saltare*, to leap) because he saw the creation of new species as happening in leaps and bounds.

But meanwhile the Dutchman Hugo de Vries had come up with a better word: 'mutations' (from the Latin *mutare*, to change).

Darwin challenged

In 1889, many years before he read Mendel, De Vries had published *Intracellular Pangenesis*, supporting Darwin's wrong hypothesis. Then, between 1901 and 1903, he published *The Mutation Theory*, with very different ideas. In it he neglected to mention that reading the *Proceedings of the Natural History Society of Brünn for 1866* had sparked his change of mind. Later, when there was an argument about who first rediscovered Mendel, he admitted he had read them … first.

Ever since 1886, De Vries had been singlemindedly working on the evening primrose, a plant that produces a spectacular number of mutations. Altogether he grew 57,000 of them, putting Mendel's 29,000 peas in the shade. And the frequent mutations of the evening primrose, he claimed, demonstrated 'mutation' to be the key to the creation of new species.

Until about 1930 Bateson, De Vries and 'mutationism' held the field. De Vries admitted that Darwin had been a great man, ingenious and hardworking, but not as great at De Vries – he always spoke of himself in the third person.

Unfortunately De Vries relied solely on the evening primrose. It is an exception, what Darwin had called a 'sport'. The species is itself a sort of huge mutation – it has triple chromosomes rather than the normal double, which make it liable to random deviation. Most normal life forms have far fewer mutations. Mutation does matter, but natural selection matters more.

Chromosomes

Chromosomes exist in the nucleus of almost every cell in the body. They are minute, wriggling, worm-like structures usually doubled in the form of a rough St Andrews cross, or perhaps a bunch of four tiny sausages. They had been discovered in 1879 when it was found possible to stain them with the new aniline dyes to make them stand out. This explains the name (from the Greek *chroma*, colour, and *soma*, body). But nobody at that time knew what chromosomes did.

In a brilliant flash of insight, Weismann had suggested that the cell nucleus could be where his germ-plasm was located, and even that it was arranged in lines along the chromosomes. Proof came in 1903 in a groundbreaking paper by the American Walter Sutton, based on a study of grasshoppers. And in it, unlike De Vries, Sutton gave full credit to the genius of Mendel.

Darwin vindicated

The rehabilitation of natural selection from about 1920 onwards was due to two outstanding men: Thomas Hunt Morgan and Sir Ronald Fisher. Morgan and his team knew from Walter Sutton's work that the chromosomes of sex cells were distinct from other cells, and they held the actual hereditary material. They were Weismann's 'germ-plasm'. Morgan had accepted 'mutationism',

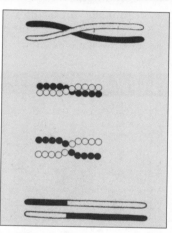

⊙ *Diagram of the crossing over of chromosomes, from Morgan's* A Critique of the Theory of Evolution, *1916*

but later came to reject the hypothesis. How this happened will be told in Chapter 11.

Fisher was a giant among statisticians. Working at Rothamsted Agricultural Research Station in Hertfordshire from 1918 onwards, he showed that the immensely complex calculations involved in cross-breeding plants still worked according to Mendel's basic laws, and they worked by natural selection with a bit of help from mutations – though Fisher did suggest that Mendel's maths was not quite as good as he thought it was, and that he may sometimes have fudged his results.

Morgan and Fisher's ideas overtook pure 'mutationism' in the 1930s. Everywhere Mendel and Darwin were accepted. Everywhere, that is except the Soviet Union.

Stalin's geneticist

Trofim Lysenko was an ardent disciple of Lamarck.

For example, he believed that if all the leaves were plucked from plants eventually the descendants of these plants would be leafless, and that if you grafted one plant on to another its descendants would acquire the charactistics of both. In 1928, he claimed to have developed an agricultural

⊙ *Trofim Lysenko in a field with wheat*

technique (vernalisation) that used humidity and low temperatures to inspire wheat to grow in spring, and that this acquired characteristic would be inherited by its offspring.

Lysenko had everything going for him as far as the Soviet leadership was concerned: his techniques replaced expensive and unavailable fertilizers, the idea that striving could permanently improve life fitted Soviet doctrine, and above all Lenin and Stalin both liked him.

From 1929 onwards geneticists who opposed his ideas were persecuted and removed from their posts, and frequently disappeared, most notably the distinguished biologist Nikolai Vavilov, who was imprisoned and starved to death in 1943. Lysenko even survived Khrushchev's denunciation of Stalin. Only in 1964 was his evidence condemned as fraudulent and by then the Soviet Union was thirty years behind everybody else. His career shows what happens when governments put political correctness before scientific objectivity.

Chapter 10

The Modern Synthesis

Population genetics

In the book that inspired Darwin, *An Essay on the Principle of Population*, Malthus had examined entire human populations and applied some

somewhat dubious mathematical assumptions. In the 1920s and '30s a new breed of geneticists returned to this overall mathematical approach. Ronald Fisher was one and J. B. S. Haldane another. They, too, worked on what was called 'Population Genetics', but used much more reliable statistical analysis. While Morgan and

⊙ *Ronald Fisher*

De Vries had examined mutations in individual specimens, these statisticians studied whole 'populations' of plants and animals.

At Rothamsted Agricultural Research Station, Fisher had data that had been collected over many years. He analysed the spread of variations using new statistical techniques that he himself invented.

In *The Genetical Theory of Natural Selection*, published in 1930, Fisher unequivocally demonstrated how Mendelian genetics and Darwinian natural selection worked in harmony.

Moths and murk

Haldane, too, delved into records dating back over a hundred years. He was looking at peppered moths. Thanks to the popularity of butterfly and moth collecting, plenty of research material was available, and he was able to demonstrate that in the 1840s light-coloured versions of these moths had predominated, only about 2 per cent being dark hued. Those with light colouring had been camouflaged when they were on pale-coloured bark and lichens on trees, whereas dark ones had been conspicuous against the light bark and thus at risk from predators.

Then the pollution resulting from the industrialisation of the Midlands killed the lichens and left the trees black with soot. Light moths continued to be born, but now they were the ones that stood out and were snapped up by predators, while the dark ones escaped notice. By 1895,

98 per cent of the peppered moth population in the Manchester area were very dark. Haldane had demonstrated that natural selection not only worked, but sometimes worked very rapidly.

Since the Clean Air Act of 1956, passed by Parliament in response to the 'great smog' that fell over London in 1952, the air has become less polluted and the moths are again lighter.

Also in the 1920s and 30s, significant progress was being made in the USA. Another statistician, Sewall Wright, worked on inbreeding in isolated populations, which he suggested could accelerate the effect of chance genetic changes, known as 'genetic drift'.

Dobzhansky

In contrast to Wright, Haldane and Fisher, the Ukrainian Theodosius Dobzhansky was a practical observer, not a mathematician. Inspired by the researches of the Russian geneticist Sergey Chetverikov (cut short by the rise of Lysenkoism), Dobzhansky had come to the USA to work with Thomas Hunt Morgan.

In 1937 Dobzhansky published *Genetics and the Origin of Species*. In it he emphasised the breadth of genetic variation in plant and animal populations and argued that natural selection was responsible, not merely for change, but, as Darwin had postulated, for the enormous diversity of animal and plant species. Natural selection enabled species to find highly specific niches. For example, there are fish in Lake Victoria in Central Africa that have become adapted to a diet solely consisting of the eyeballs of other fish, while another closely related type of fish is adapted to filter plankton, and others to snap up insects or scrape off fish-scales.

Wright's and Dobzhansky's conclusions are a vindication of those Darwin had reached from studying the range of plants and animals of the Galapagos Islands.

Dobzhansky showed diversity to be a direct result of evolution: insects have been around for 345 million years, and this has given them time to become very specialised – hence there are 950,000 known species. Bacteria have been here a lot longer, and no one knows how many species there are. A recent survey by Jason Gans of the Los Alamos National Laboratory found about a million bacterial species in a single gram of soil.

The fossil record revised

Bacteria, though highly successful, are very simple in structure. As evolution moved on to more complex life forms, various distinct body plans emerged. Some proved more useful than others and therefore survived – in particular vertebrates, with their backbones and skulls, and insects, with their external skeleton and division into head, thorax and abdomen. Each of these plans in time diversified in adapting further to the range of environments in which individuals thrive.

Fossils might have been expected to record this adaptation, but it takes very special circumstances to preserve a fossil. In a few instances insects became trapped in amber; sometimes a particular creature or plant left its mark in mud that later turned to slate, and very occasionally chemical reactions created exact copies of extinct animals' bone structure in stone. *Tyrannosaurus rex* must be the world's favourite dinosaur, yet to this day only about twenty, all incomplete, skeletons have been found, and five of those were in a single group that turned up in Montana in 2000.

⊙ *Cast of an* Archaeopteryx *fossil, Geneva Natural History Museum*

By Darwin's day fossil investigation had made clear that life on Earth had existed for a very long time and had changed a lot over the years. But there were huge gaps in the fossil record. Worse, none of the tiny proportion of creatures preserved as fossils really shed any light on natural selection since none, not even *Archaeopteryx* or *Ichthyosaurus*, could illustrate the gradual development of new creatures through slow adaptation over numerous generations.

In consequence, many palaeontologists felt the fossil evidence supported not Darwin's ideas, but those of Richard Owen or later those of De Vries. Fossils seemed to demonstrate a series of leaps forward into ever improving life forms: all leading onward and upward to nature's final triumph: us.

Until 1944, that is. In that year the American George Gaylord Simpson, arguably the greatest palaeontologist of the twentieth century, demonstrated that in reality the fossil record is consistent with the 'irregular, branching, and non-directional pattern' predicted by the 'survival of the fittest'. Further discoveries over the last sixty years have endorsed this assessment. Darwin has, once again, been vindicated.

Keeping fit

'Fitness' is a word that has changed its meaning. When he coined the phrase 'the survival of the fittest', Herbert Spencer did not mean that the healthiest and strongest animals would survive. He meant that those organisms that survived would be the ones that were best fitted to their particular environments. And 'fitting in' generally required flexibility and adaptability – as Darwin wrote:

'It is not the strongest of the species that survives, nor the most intelligent that survives. It is the one that is the most adaptable to change.'
As usual, Darwin was right, and this fact has

had some curious consequences. Fish live in the sea, but some resourceful fish came out of the sea. Most died or went back, but a few eventually adapted, survived and evolved over millions of years to become land-based reptiles, and then mammals, and so on. Much, much later a few recidivist mammals returned to the sea, adapted and thrived: whales and dolphins live in the sea because it suits them. They look like fish because fish-style streamlining works best in water, but they remain mammals. This process, the opposite of divergence, is 'convergent evolution': animals of very different ancestry become similar because they have to solve the same problems and live in the same environment.

Birds, bats and insects

Examine the skeleton of a bird. It resembles a dinosaur because birds developed fairly directly from some types of dinosaur. Look at the skeleton of a bat, and it resembles that of a mouse because it is a mouse –

⊙ *Brown big-eared bat. Illustration from* Brehms Tierleben, *1927*

but one equipped with wings and radar. Insects have wings too, but these developed in a completely different way from those of birds and bats. Birds and bats lost their forelegs, insects kept all six legs. And as for penguins! – They are birds that think they are fish, and thrive in the coldest, most desolate parts of the world. Fishes, birds (penguins), and mammals (dolphins and seals) all have fins or flippers because that is the best way to move fast in water and get away from predators.

All these examples of convergent evolution demonstrate 'the survival of the fittest'.

Leftovers and old-timers

Most creatures' bodies are adapted to their habitat, but many retain features left over from a previous incarnation and previous habitats, such as our tiny hidden tail – the coccyx – or the python's pelvis. They do not especially inhibit us, so they remain in place. On the other hand, some species have hardly bothered to adapt at all.

Natural selection will not produce novelty just for the sake of it. And although we so-called 'higher animals' are very complex, complexity is not an end in itself either. Novelty and its attendant complexity appear only if they are needed. Alongside new-fangled contrivances like us, many highly successful, simple, old forms live on – organisms that are well adapted to their environment and have never had any need to update themselves.

Sponges are simple life forms ideally suited to their habitat, and for millions of years natural selection has left them almost completely alone. Anemones, sea urchins and jellyfish are ancient, simple in structure and survive in massive numbers.

All of these quaint anomalies would seem absurd if animals were constantly improving themselves or being improved by a tinkering Creator. But they make absolute sense if the motor for change is not self-improvement or perfectionism but simple survival.

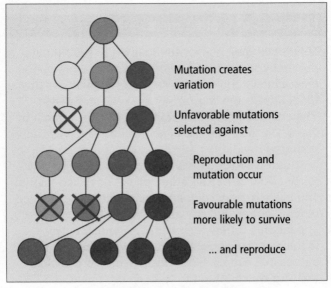

Mutation creates variation

Unfavorable mutations selected against

Reproduction and mutation occur

Favourable mutations more likely to survive

... and reproduce

⊙ *Natural selection of a population for dark coloration*

The modern synthesis

In 1942, the biologist Julian Huxley, Thomas's grandson, summed up these ideas in *Evolution: the Modern Synthesis*. Evolution, he said, is predominantly the product of small, gradual, genetic changes working in tandem with natural selection, and to a lesser extent genetic drift and genetic flow – the way sexually reproducing creatures often end up moving, or being moved, away from home to breed. Variations arise by chance. They can cause change immediately, remain dormant forever, waste away, disappear completely, or weigh in when altered circumstances make them of value to survival. Minor advantages come to be crucial over many generations. But natural selection is not a matter of chance; it is an iron law. And it works through whole populations. This Darwinian truth was to lead Julian Huxley and others seriously astray in the 1930s.

Chapter 11

From Eugenics to DNA

The dark side

Although the 1930s and '40s saw the revival of belief in natural selection, in another respect these years were a bad time for Darwinism. Charles Darwin had been concerned that civilised society enabled weak humans to survive and breed, and that this weakened the general stock. His cousin

⊙ *Sir Francis Galton, 1912*

Francis Galton took up this theme in 1883 and named it 'Eugenics': the 'science' of ensuring that the healthiest and best humans breed with each other and the sickly and degenerate do not breed at all. It was supported by left-wing humanitarians such as H. G. Wells, Bernard Shaw, Julian Huxley himself and Woodrow Wilson, and even by the statistical genius Ronald Fisher. Yet it was a cruel, scientifically flawed concept that led to the sterilisation of thousands of citizens of the United

States, Belgium, Brazil, Canada and Sweden. Worse, in Nazi Germany, Darwinian eugenics was used to justify the mass extermination of Jews, gypsies (Roma), homosexuals and the disabled. After the Second World War a new word was added to 'gene' and genetics', the international crime of 'genocide'.

But while eugenics was a dangerous deviation, genetics thrived on the firm foundations of Mendel and Darwin. Between Darwin's death in 1882 and the start of the First World War a whole vocabulary was created.

A new vocabulary

Genes are Mendel's 'factors', in which characteristics are passed from one generation to the next. For example, a plant with red flowers must carry a gene for that colour. But, as Mendel showed, a colour may occur in one of two forms. These two alternatives are called the 'alleles' of that gene. For example, the colour gene in pea plants can occur in the allele for a white flower or the allele for a red flower.

Two other words the geneticists came up with are 'heterozygous' and 'homozygous' (from the Greek *hetero*, different, and *homo*, same). The term 'heterozygous' is used of a plant when two alleles are different, such as red and white, and 'homozygous' when two alleles are the same, such as red and red.

Chapter 6 described what happened when Mendel fertilized the female plant with pollen from the male plant. This is the same process in the new scientific jargon:

1. Sex, or reproductive, cells are 'gametes' (this is Mendel's word; Weisman's 'germ-plasm' fell into disuse).
2. The gametes from the male plant combine with the gametes of the female plant.
3. They produce a fertilized egg or seed called a 'zygote'. A zygote contains genetic information from both parents.
4. If a zygote contains one allele for white flowers and one allele for red flowers, the resultant hybrid pea plant will be heterozygous, since it has two different alleles.
5. If the zygote contains a gene with two red alleles, it is homozygous.
6. In both cases the plant would be red, because red is dominant.
7. White is recessive. Only if both alleles are white will the flower be white.

At base this is a simple concept, and more or less the same system works for us and for every other animal and plant.

Morgan's flies

The American zoologist Thomas Hunt Morgan was mentioned above, in Chapter 9, as one of those responsible for reviving Darwin's ideas. Morgan did what Darwin did with pigeons and Mendel did with peas, only using fruit flies. Mendel had to wait six months for each new generation; fruit flies live short lives, lasting just fourteen days. Hence they produce a rapid sequence of generations. Morgan and his team, which included Dobzhansky, bred millions of them and at first, accepting De Vries' hypothesis, did all they could to create mutants using X-rays, acids and anything else they could think of.

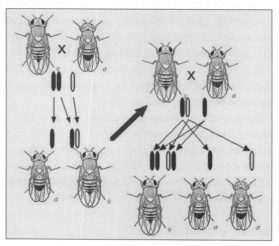

⊙ *White mutation and sex-linked inheritance in fruit flies.*
From Morgan's The Physical Basis of Heredity, *1919*

Every fly had red eyes, until suddenly one – a natural mutant, not one of the grotesque monsters produced by the X-rays or other means – appeared with white eyes. Morgan mated it to an ordinary red-eyed fly. The first generation produced 1,234 red-eyed flies and three white, but white-eyed flies appeared in larger numbers from the second generation onwards. It was Mendel's red and white flowers all over again, with the red dominant and the white recessive. Morgan realised that this was how evolution happened – through tiny mutations working in tandem with natural selection.

Chromosomes again

Walter Sutton had proved that genes were carried on chromosomes. Improved microscopes showed there to be a regular number of chromosomes per cell: fruit flies have twelve, hedgehogs eighty-eight and humans forty-six – though our sex cells (Mendel's gametes) have only twenty-three.

All our body cells are constantly dying and replacing themselves, and changes occur, for instance as we grow older. But as Weismann correctly pointed out, the gametes are uninfluenced by what happens to the rest of our bodies. A man may lose a leg and his hair may go white, but if he then fathers a child it will not consequently have white hair and only one leg.

Like Mendel's peas, children get half their chromosomes from one parent and half from the other, through their parents' gametes. The complicated process that turns the gametes of the two parents into a new, unique, individual human being would require another entire book. However, all humans have vastly more in common with each other than the minor differences we make such a fuss about.

For our present purposes we need to look at why, when a man and a woman have a baby, it doesn't turn out to be a fruit fly or a hippopotamus, but is none the less the product of evolution from a very different creature.

Morgan knew the genes carried by sex-cell chromosomes stored the template for building a new individual, and that genes from the two parents were passed from generation to generation. Each chromosome held numerous genes, and each gene held the code for a different aspect of the new individual: how many legs it had, what type of eyes, and so on.

Genes again

Genes became more specific: they were now seen as segments of chromosome. Each minute segment carried a tiny bit of specific information about an organism: one gene dictated whether Mendel's peas were tall or short, another what colour their flowers were, another what shape the seeds were and others encapsulated many thousands of explicit directives.

In the sex (or germ) cells the genes on the chromosomes determined what the offspring would be like: not just whether the new fly's eyes or the new flower's petals would be red or white, or what sex it would be, but whether it would grow into a pea or delphinium, whether indeed it was to be a plant or an animal.

Morgan recognised that, as Weismann had surmised, the genes appeared in sequence on the chromosomes, one after the other, and as early as 1911 he had mapped the position of five genes on the fruit flies' chromosomes. Ten years or so later he had mapped the position of more than two thousand genes on fly chromosomes.

Mutations again

It became clear that genes do not always come in two straightforward versions, one dominant and one recessive. There can be many different versions of the same gene, many different alleles. A mutant allele can occasionally cause a marked change to an organism. However, since most traits depend on a range of different genes working together, mutating one of them normally results in only a tiny change, or, more usually, no change whatsoever. A major mutation will normally stop the whole process from working, and lead to death.

So De Vries was right in that mutations have a part to play. but Darwin, who accepted mutations (he called them variations), more accurately described evolution as a combination of minor mutation and the slow process of natural selection, in which natural selection predominated.

By the early twentieth century it was clear that both processes worked through the action of genes; but that raised the question of what genes were made of.

Chemical analysis demonstrated that chromosomes contain various different proteins and nucleic acids (acids in the nucleus). Somehow they had to carry the genetic information to the offspring. By 1944 the Anglo-American bacteriologist Oswald Avery had shown that this 'transforming principle' was one of the acids in the cell's nucleus. It was deoxyribonucleic acid, or DNA.

The four bases

But what was DNA made of? The Russian-born chemist P. A. T. Levine worked with Avery at the Rockefeller Institute, but they did not get on well. Levine was convinced that DNA had nothing to do with genetic information. He felt DNA had much too simple a structure, since it was composed of just four chemicals: adenine, guanine, cytosine and thymine. These chemicals are called the 'bases', and are now generally shortened to A, G, T and C. And Avery was right – they are the 'transforming principle'.

Later, Levine discovered that each base is stuck to a big sugar molecule (deoxyribose). These sugar molecules are all strung together along the chromosome, linked like a pearl necklace. The 'string' is a phosphate called phosphodiester.

Phosphodiester ... O^C ... O^T ... O^T ... O^A ... O^G ... O^C ... O^G

Base Base Base Base Base Base Base (above)
Sugar Sugar Sugar Sugar Sugar Sugar Sugar (below)

If there were differing amounts of these four bases in different animals and plants, that would explain a great deal. But, as far as could be observed at the time, DNA always contained equal quantities of each.

However, the order in which these four bases are arranged along the 'necklace' varies. Part of a string might be:

A. T. A. C. C. C. A. T. G. C. G. A. G. A. G. T. A. T,
and so on.

Even more significantly, the order in which the bases are arranged differs from species to species.

Chargaff's rules

Then superior techniques revealed that the percentages of each base were not, after all, exactly equal. Chargaff's rules are named after Erwin Chargaff, a Jewish biochemist born in Chernivtsi (then part of the Austro-Hungarian Empire and now in Ukraine), who fled Europe for the USA in 1935. His rules, elaborated in 1949, state that:

1. In all animals and plants the percentage of Adenine in DNA is always virtually equal to the percentage of Thymine, and the percentage of

Cytosine is always virtually equal to the percentage of Guanine.

2. And, after all, different species do have different percentages of these pairs of chemicals.

Cracking the code

Levine and Avery had demonstrated that the order in which the bases are strung out differs in different species. Chargaff then showed that the amount of A, G, T and C on chromosomes differs in different species. In the following examples, all figures are approximate:

- A human being has roughly 30 per cent each of Adenine and Thymine, and 20 per cent each of Cytosine and Guanine.
- An octopus has roughly 33 per cent each of A and T, and 17 per cent of C and G.
- Wheat has roughly 27 per cent each of A and T, and 23 per cent each of C and G.

These variations are the key to the differentiation of species: each separate species has different amounts of A+T and of C+G and they are strung out in a different order; so each has a separate 'genetic code'. The genetic code dictates whether any organism is a pea, a whale or a fruit fly.

That just left one problem. How on earth did it all work?

Chapter 12

The Secret of Life?

Loud laughter

When I was at Cambridge in the mid-1960s, I knew a number of geneticists, and we often had lunch at a rambling old inn called the Eagle. They would sometimes wave or chat to a flamboyant figure sitting over his pint at the bar, surrounded by admirers and frequently bursting into loud laughter. To me he was just another Cambridge character. To the world he was the front half of Crick and Watson, the discoverers of 'the Secret of Life'.

Crick and Watson

Francis Crick and James Watson were a pair of mavericks. Crick, educated at a minor English public school, was twelve years older than the American James Watson. He had a degree in physics, though not a particularly good one, from London University, and his attempt to get a PhD was interrupted by the war when, to his delight, a bomb destroyed his research equipment. He spent the rest of the conflict in the Admiralty, developing magnetic mines. With the coming of peace he became interested in the physics of biology. In

1949, aged thirty-three, Crick arrived at the famous Cavendish Laboratory in Cambridge, where Sir Laurence Bragg, the conventional English gentleman in charge, had a strong antipathy towards him. When Francis Crick met James Watson in 1951 he still had comparatively little knowledge of biology but an amazing talent for coming up with brilliant and inspired ideas, some of which were right.

Watson had been a boy wonder – the star of quiz shows and a university graduate by the age of nineteen. He had no interest in chemistry, but was nevertheless keen to study genes. In his old age Watson was described as 'brash, bumptious, and brilliant', and at twenty-three he was infinitely more so. He and Crick got on well. Their skills

⊙ *James D. Watson*

complemented each other, they shared an interest in warm English beer, and they also shared an arrogant confidence in their ability to discover the structure of DNA and how it passed on inherited characteristics. 'I have never seen Francis Crick in a modest mood,' was to be the first sentence of Watson's account of the next two years.

The race was on

They weren't the only ones in the field, and no bookmaker would have made them favourites to win the race; in fact they would have been 100–1 outsiders. The favourite would probably have been the American Linus Pauling, the greatest living chemist, who had virtually founded the science of molecular biology. He had suggested that the layout of the proteins in cells could be in the form of a spiral: a corkscrew shape technically called a helix (Latin *helix*, a spiral). Could some form of helix also be the structure of the nucleic acids? Pauling thought this was possible. Another major contender would be Erwin Chargaff, of Chargaff's rules fame, though he did not accept the spiral idea. A third was Maurice Wilkins, a New Zealander working at King's College, London. Wilkins had the advantage of a very superior piece of equipment: a camera capable of high-resolution X-ray diffraction, which could take X-ray images of DNA. He also had a very superior person working with it: the brilliant, twenty-nine-year-old Rosalind

Franklin. She would have been the bookmaker's fourth favourite.

Wilkins and Franklin

Franklin and Wilkins hated each other as much as Crick and Watson liked each other. But Crick and Wilkins also got on well, and Wilkins was happy to share his team's findings with the pair at the Cavendish. While Franklin and Wilkins took photographs, Watson and Crick followed Pauling's example and made a model – a wonderfully complex model that involved three intertwined helixes with the bases stuck to the outer side. Which was quite wrong.

Ignorance is bliss

Later, they came to accept Rosalind Franklin's idea that the bases were on the inside of the helixes. The bases faced each other. Or not quite: the base molecules were flat discs, and Crick guessed that they actually slotted together, like packs of cards shuffled together. And the bases of one helix stuck to the bases on the other. But it still involved three intermeshed corkscrews – which did not leave much room for the bases themselves.

Chargaff had shown that the number of A and T bases equalled each other, as did the number of C and G bases. This might have given the Cambridge pair a clue, except that Watson didn't believe

Chargaff, and Francis Crick was the one person working on DNA who had never heard of his rules.

Un-American activities

Things were hotting up: Linus Pauling was on the same track, and was coming over to London to see Wilkins. A mind as good as his would instantly spot the answer from the King's College photographs. Then politics intervened. The USA was in the throes of McCarthyism and the Un-American Activities Committee. Pauling was an advocate of World Peace (he was to be awarded a Nobel Peace Prize to add to his Nobel Chemistry Prize), and as such was obviously a very dangerous man. He was not allowed to board his plane.

Chargaff in a huff

In 1952 Chargaff turned up in Cambridge. Watson used a mutual contact to wangle an introduction. Chargaff had no idea who Watson was, and was amazed to find a crazy, fuzzy-haired, ale-drinking American who had developed a fake English accent (Watson was convinced the accent and the haircut helped him pick up Scandinavian au pair girls), accompanied by a loudmouthed Englishman who, though he had now heard of the famous rules, could not remember them at all clearly. Chargaff went off in a huff, refusing to listen to anything the pair had to say. But his rules were to prove invaluable to them.

Pauling goes wrong

Late in 1952, Linus Pauling's son Peter was in Cambridge, and chanced to make friends with Watson. They shared a hobby: the Scandinavian au pairs. The professor was himself writing a paper outlining the structure of DNA. Peter wrote to his father:

'You know how children are threatened, "You had better be good or the bad ogre will come get you." Well, for more than a year Francis and others have been saying to the nucleic acid people at King's, "You had better work hard or Pauling will get interested in nucleic acids." I would appreciate very much a copy of your article.'

The advance copy arrived. Peter passed it to Watson and Crick. Fearful of being pipped at the post, they opened it in dread – then read it in amazement. Like them, the world's greatest chemist had three helixes, but had made the same mistake they had made. He had stuck the bases on the outside. Watson checked and rechecked the figures, and there could be no doubt about it: Pauling had made several fundamental mistakes. His version would have fallen to pieces.

Franklin falters

Meanwhile, Rosalind Franklin had been taking better pictures of a better form of DNA (the B version) with a better machine. As a result, she was beginning to doubt that DNA was a helix. In a rare moment of friendship she lent one of the photographs to Wilkins: picture number 51. He showed it to Watson. In Watson's words, 'The instant I saw the picture, my mouth fell open and my pulse began to race.'

At a glance he saw that Rosalind Franklin was wrong, in that DNA was a helix. But it was probably not a triple helix. Back in Cambridge, Watson and Crick worked feverishly and came up with a double helix: a kind of spiral staircase or, more accurately, a twisted ladder. The bases were linked together to form the rungs.

The final furlong

It was now 1953, the year of the Queen's coronation, the year a British team conquered Everest and the year Roger Bannister ran the first four-minute mile. On 21 February Watson and Crick had a sudden key insight, inspired by Chargaff's rules and the calculations of a bright young mathematician called John Griffiths.

Technically the A and G bases are called purines, and the T and C are called pyrmidines. Purines and pyrmidenes are not the same length, and Crick and

Watson could not jiggle them around to get linked pairs to fit inside the diameter of the helix – which Franklin's pictures had enabled them to calculate. Then they did some more sums (Crick fiddled around with cardboard cut-outs) and worked out that when A and T were stuck to each other they added up to exactly the same length as the combination of C and G, and would fit neatly into the space available. What is more, with the bases paired in this way and DNA being a double helix it would have a good, solid construction. Each rung of the twisted ladder would be the same length.

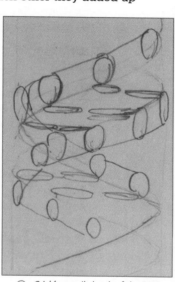

⊙ *Crick's pencil sketch of the DNA double helix, 1953*

Crick's reaction was to rush into the Eagle and loudly tell the world, 'We've discovered the secret of life'.

Considerable biological interest

Watson and Crick sent an advance copy of their paper to Maurice Wilkins. Wilkins replied on

18 March, making some minor emendations and commenting, 'I think you're a couple of old rogues, but you may well have something.' The letter ends 'as one rat to another, good racing', and is now in the Crick archive at the University of California, San Diego.

Wilkins was right: they did have something. On 25 April Crick and Watson published a nine-hundred-word paper in *Nature* entitled 'A Structure for Deoxyribose Nucleic Acid'. It began:

'We wish to suggest a structure for the salt of deoxyribose nucleic acid (D.N.A.). This structure has novel features which are of considerable biological interest ...'

And to their credit, it ended:

'We have also been stimulated by a knowledge of the general nature of the unpublished experimental results and ideas of Dr. M. H. F. Wilkins, Dr. R. E. Franklin and their co-workers at King's College, London.

Reproduction

First, a disclaimer. Reproduction is a highly complex process; what follows is an extreme simplification and specifically related to evolution,

Our body cells are constantly dying and being replaced by almost exact copies. Crick had worked out how DNA replicated itself before *Nature* published the paper he wrote with Watson.

When cells reproduce themselves the two helixes rip apart, like a zip unzipping. Each side of the zip sets about building itself a new opposite side to make up for the one torn away. And this copies the one ripped away. Because:

Base A could only pair with T
Base G could only pair with C
Base T could only pair with A
Base C could only pair with G.

In practice, the copying is not always perfect: DNA gets damaged by, for example, smoking and excessive sunlight, and with only two different pairs of bases the matching and sequencing can sometimes get muddled. Minor mutations occur that multiply over the years; cells die and are not replaced, leading to the bodily changes of aging and the deadly distortions of cancers. Weismann's 'barrier' between body cells and sex cells means that none of this is directly related to evolution. However ...

Sexual Reproduction

In their brief paper in *Nature*, Crick and Watson stated, 'It has not escaped our notice that the specific pairing we have postulated immediately suggests a possible copying mechanism for the genetic material'. And this proved to be the case.

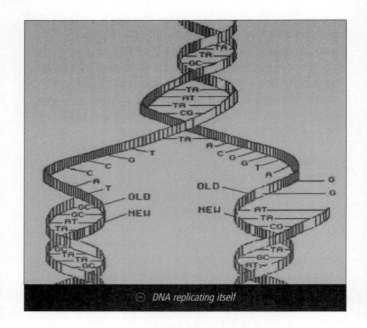

⊙ *DNA replicating itself*

The key requirements for genetic material are that it must be able to carry information; that information must be able to be copied; and (occasionally) the copies must be a little different from the original. And DNA is capable of all three requirements: in sexual reproduction the basic technique of replication also applies, and similarly the replication is not always perfect, creating occasional mutations.

When the male gamete fertilizes the female gamete, each parent's DNA replicates itself, creating two distinct sets of chromosomes. These then combine together and reduce down in two stages to produce a single set of chromosomes carrying a mixture of DNA from both parents. This is the zygote, which, in animals, will develop into the embryo.

Meiosis and polymorphism

This process is called 'meiosis' (Greek, *meiosis*, 'getting smaller'). Meiosis means that over many generations millions of individuals' DNA becomes intermixed.

Because a zygote contains a virtual copy of the two parents' genes, offspring tend to resemble their parents, though Mendel's laws require that some characteristics are dominant and others recessive, and that some children look more like their great-grandmother than their mother.

Over the millennia, as generation after generation mated with outsiders rather than their own brothers and sisters, we have each acquired a fine medley of genes, producing diversity without the need for frequent mutations. Today one individual human differs from another by, typically, one DNA base in 1,000, and we each have an individual and probably unique DNA map.

This diversity – the way each individual human, plant or animal has a slightly different form, is called 'polymorphism' (Greek, *poly*, many, *morph*, form). Polymorphism makes natural selection feasible.

The Nobel Prize

James Watson, still only twenty-five in 1953, got a top job at Harvard, published a seminal textbook and wrote his own version of the events in *The Double Helix,* which hardly does justice to Rosalind Franklin but is still a classic of scientific autobiography. Happily, when Crick and Watson got their Nobel Prize in 1962 they shared it with Maurice Wilkins. Poor Rosalind Franklin, who deserved it as much as anyone, had died of ovarian cancer in 1958.

In the beginning was the word

For twenty-two years Crick went on working in Cambridge, drinking beer in the Eagle and doing brilliant research. By 1961 he and his team had worked out that the bases A, T, C and G actually function in three-letter groups or 'words': CTT, GTA, ACG, etc. These words are called 'codons', and there are only sixty-four of them.

Each codon orders up a particular 'amino acid'. There are twenty basic amino acids, and they do the real work. Like children's variously shaped wooden bricks, they can fit together in many ways, forming a wide variety of proteins which will build flesh, blood and bone or, for that matter, roots, sap, leaves, and so on.

But why should it take sixty-four codons to produce twenty amino acids? No less than six different codons (CTT, CTC, CTA, CTG, TTA, TTG) all produce exactly the same amino acid, called Leucine, and another six all produce Serine, and those are only two examples. It works in practice, but it hardly has the subtle elegance one would expect of 'intelligent design'. It is clumsy and makeshift – a product of the system of trial and error Darwin called 'natural selection'.

Genetic engineering and the human genome

Much has happened since the 1960s.

Genes are made up of sequences of DNA. By 1969 the first complete gene had been isolated, and 1970 saw the construction of an artificial gene. In so-called 'genetic engineering', scientists can now build and manipulate very long DNA sequences, which are then introduced into cells supposedly to 'improve' food products or correct faults.

Every cell carries an animal or plant's entire DNA map, its 'genome'. This can now be replicated (or 'cloned') to produce an almost exact copy. In 1996 Scottish scientists cloned the first mammal: Dolly the sheep.

In 1988 Watson set up a mammoth international project to map the human genome, which contains over three billion A, T, C and G base pairs. Chromosome Number 1, the longest human chromosome, has 247 million of them. Watson guessed they would find 100,000 genes and they would finish in 2150. They found 25,000, and completed in 2003 – but he and Crick did sometimes make really big miscalculations.

In 1977 Crick moved to America, where he continued to do important research and wrote a book suggesting that life originated from outer space. He died in 2004.

⊙ *Francis Crick, 2004*

Junk genes

To say that scientists now understand the human genome is not wholly true: they know about the parts involved in evolution and reproduction. But this still leaves a huge quantity of 'junk' genes: a vast rubbish tip of dead and discarded genes, so-called 'pseudogenes', which make up over 90 per cent of the genome, and which may or may not have a purpose. Research is beginning to investigate these cast-offs. They include, for example, those genes that once gave us a far better sense of smell and bodies covered in hair. This discarded 'junk' makes it clear that evolution is far from being the product of 'blind chance', but is more a process of trial and error, or natural selection.

Darwin was Right

In 2006, research in Germany managed to analyse the DNA of a Neanderthal skeleton. We do not share its genome. Huxley and Darwin were right. Our ancestor, Cro-Magnon man, came out of Africa about 100,000 BCE, and displaced the Neanderthal through 'survival of the fittest'.

Family trees

DNA is again and again providing the hard evidence Darwin lacked. Genes are inherited from ancestors, so the best evidence of inheritance from an extinct ancestor comes from gene sequences. When you analyse DNA, it soon becomes clear that apparently different species can contain a high proportion of the same genes. They must have inherited these from a common ancestor (especially if they include mutations than are unique to that ancestor). If we have the same genes, then at some time in the dim and distant past we had the the same ancestor. This is what Darwin called our 'common descent' – once again, he was right.

Absurd in the highest degree

And then there is 'the eye', Dean Paley's star exhibit to prove 'intelligent design'. Here's what Darwin said about it:

'To suppose that the eye, with all its inimitable contrivances for adjusting the focus to different distances, for admitting different amounts of light, and for the correction of spherical and chromatic aberration, could have been formed by natural selection, seems, I freely confess, absurd in the highest possible degree. Yet reason tells me that if numerous gradations from a perfect and complex eye to one very imperfect and simple, each grade being useful to its possessor, can be shown to exist; if further, the eye does vary ever so slightly, and the variations be inherited, which is certainly the case; and if variation or modification in the organ be ever useful to an animal under changing conditions of life, *then the difficulty of believing that a perfect and complex eye could be formed by natural selection, though insuperable by our imagination, can hardly be considered real.'*

One hundred and thirty-five years later he was proved right. In 1994 computer modelling demonstrated that a complex eye could evolve from a simple light-sensitive patch in 400,000 steps. And although our eyes are very different from, say, a fruit fly's, in that same year, 1994, fruit fly eyes and human eyes were in fact shown to share a common group of genes. It has since been found that all sighted creature have the same set of light-sensing proteins (opsins). This means they all have a shared point of origin, a point from which our various eyes have, little by little, evolved. And that evolution can be mapped in our genes.

The more closely two species are related, the more their gene sequences will have in common. This can produce some surprising results. Whales are sea-going mammals, but what can be their closest relatives? The answer turns out to be the water-loving hippopotamus. By contrast, it is pretty obvious who our closest cousin is.

Apes and men

You only have to compare our skeletons with those of apes to see how closely we are related. And the skeleton of oldest form of the genus *Homo*, *Homo habilis* (Latin, handy man), was even more ape-like. So what is the link between humans and apes? Like lemurs, monkeys and marmosets, both apes and humans are primates – the word reflects the outlook of Erasmus Darwin and many others, since in Latin it means 'principal', 'first rank'. Apes and humans are not just mammals; we are the top mammals because we have highly developed hands (and feet, in the case of many apes) and big brains. As Huxley suggested, the apes most closely related to us are the chimpanzees. Mankind used to be categorised as unique in being a toolmaker, but David Attenborough has pointed out that chimps can also make primitive tools. He observed how, on the way to raid a termite hill, they will break off a twig, trim it to size and then stick it in the hill, where it will become covered in gallant soldier termites fighting to preserve the sanctity of their home – who then get eaten for their pains.

Hominini and panini

Charles Darwin seems to have been right in surmising that mankind and the apes both originated in Africa. While it has long been accepted that Darwin and Huxley were correct in thinking that humans and the apes have a common ancestry, between the 1920s and the 1960s experts held that the nineteenth-century biologists had been mistaken in postulating a close connection, believing that primitive humans had split away from the apes well before the separate types of ape emerged. Modern DNA evidence supports the

Victorian thesis: we all marched along together until the gorillas went their own way seven million years ago and about five or six million years ago the chimpanzees and 'hominins' parted company. 'Hominins', or 'hominini', is the new word for humans, their ancestors and relatives; the gorillas are 'gorillini' and the chimps, somewhat surprisingly, 'panini'. Hominins include *Homo sapiens* ('Intelligent Man', as we rather complacently call ourselves), also the Neanderthals, *Homo erectus*, Peking Man. Since the 1970s excavations in Africa have revealed several other close relatives, with smaller brains, who actually lived alongside us for a while and then died out, just as Darwin would have expected.

Chaps, chimps and carrots

It is often stated that we share 99 per cent of our DNA with chimpanzees, which seems overwhelming until you realise that this means something like forty million differences in the four nucleic acid bases. To put this in some perspective, we also share 40 per cent of our DNA with a carrot and 50 per cent with the mustard weed plant. As part of the Human Genome project, the University of Washington spent $30 million on 'Chimp Sequencing and Analysis'. Among the things that they learnt is that chimps rarely get cancer and never get Alzheimer's, though they do get diabetes. So, though five million years is not a long time

compared with the hundreds of millions of years
that living things have existed, there have been
diverse developments. DNA research has
established that quite a lot has happened to
separate people and apes. According to the
University of Washington study:

> 'While there's only been a 1 percent change in
> the shared genes of chimps and humans in 6
> million years, each species has also either added
> or dropped another 1.5 per cent of their genes.'

In practice we are now about 96 per cent
genetically the same. Among the changes for us
humans was the evolution of our genetic code for
speech. On the other hand we lost, or 'switched off'
thirty-six genes related to our sense of smell and
one gene related to hairiness – that was only about
240,000 years ago, but the effects were significant.
So Darwin and Huxley were right: we are closest to
chimpanzees, but not so close that, as someone
remarked, 'we feel we can invite each other to our
respective tea parties'.

As for whether primates truly are the most
successful creatures on Earth – really are 'prime' –
we certainly do not come first in terms of age, or
numbers, or the ability to survive adverse
conditions. To quote Richard Dawkins:

> 'For the first half of geological time our ancestors
> were bacteria. Most creatures still are bacteria,

and each one of our trillions of cells is a colony of bacteria.'

Bacteria survived incredible changes in atmosphere and temperature that would have killed us in microseconds. Our evolution away from these tough little survivors could yet prove to have been a big mistake.

Another primate

The word 'primate' has another meaning, one used at the beginning of this book. It can also mean the chief bishop of a country. The husband of Cecil Alexander, the author of 'All things Bright and Beautiful', rose to be Archbishop of Armagh, Primate of all Ireland, from 1896 to 1911. One of his predecessors was Archbishop James Ussher, who was Primate of Ireland from 1625 to 1656. He is important to us because it was Archbishop Ussher who worked out the age of the world. In his *Annales veteris testamenti, a prima mundi origine deducti* ('Annals of the Old Testament, deduced from the first origins of the world'), he came up with a pretty exact answer. The world, the archbishop tells us, was created at nightfall prior to Sunday, 23 October, 4004 BCE.

It is easy to mock, but his calculation was a remarkable work of scholarship – Isaac Newton would try to do the same thing and fail. Ussher needed to coordinate and cross-reference Persian,

Greek, Latin and Hebrew texts, which required a thorough knowledge of those languages. He got the death of Julius Caesar right (44 BCE) and that of Alexander the Great (323 BCE). He remained more or less on track all the way back to Nebuchadnezzar, but from then on he relied on the Hebrew Torah and Genesis for a highly speculative sequence of generations stretching back to the Creation. Ussher, like modern scientists, had to work with the data available to him; and scientists, as Darwin was aware, are only as good as their data. It was a beautiful bit of work, but as Thomas Huxley was to comment, 'The great tragedy of science is the slaying of a beautiful hypothesis by an ugly fact'.

Darwin's legacy

More than any other book, *On the Origin of Species* established the primacy of the scientific method founded on hard, sometimes ugly facts. Darwin summed it up in two sentences:

'Science consists in grouping facts so that general laws or conclusions may be drawn from them',

and

'I am a sort of machine for observing facts & grinding out conclusions'.

The scientific method calls for rigorous objectivity. Before the publication of *The Origin*

Richard Owen, a committed Christian, took God for granted and attempted to make the observed facts of palaeontology fit his beliefs. After its publication August Weismann's objectivity led him to cut the tails off thousands of mice simply to test whether his own hypothesis might be wrong.

Before Darwin, a Creator God was the most satisfying and logical explanation of the complexity of Life. Darwin did not disprove God's existence – he simply made it unnecessary. Darwin showed there was a perfectly feasible alternative explanation, and generations of scientists have since been refining and expanding his ideas, filling in gaps and frequently proving that he seems to have been right.

And because non-scientists could understand Darwin his influence spread far beyond the scientific community.

Judaeo-Christian and Islamic thinkers had all agreed than 'Man' was created in God's image. Darwin tells us, 'Man' is no more than 'a more highly organized form or modification of a pre-existent mammal.' We are the product of natural selection, survivors of a merciless war of attrition. We have survived so far because we have adapted well to our environment, not because we are the highest form of life. To quote Darwin again, 'It is not the strongest ... species that survives, nor the most intelligent. It is the one that is the most adaptable to change.' Whales, barnacles and bacteria have also won through in this deadly obstacle race. Come the next major natural catastrophe, they could well prove to be better survivors than humans.

Modern 'intelligent design' is a less profound concept than the old 'teleological argument', which implied a purpose to life. Without a God, or gods, is mere survival an adequate purpose? Can it provide an alternative justification for morality and ethics – the golden rules that have produced so much that is best in human nature, art and society? Darwin postulated the good of the tribe or the good of the species as alternatives, but modern research has ruled these out. Are we then, like soldier termites, simply the expendable slaves of our genes?

Charles Darwin was a kindly, humanitarian man, a loving father and husband, a man sickened by the sight of pain, a hater of slavery, but as he grew older he came to feel he could not deny the 'ugly fact' that, in his own words:

> 'The universe we observe has precisely the properties we should expect if there is, at bottom, no design, no purpose, no evil, no good, nothing but blind, pitiless indifference.'

The breath of God

Such was the considered opinion of Darwin in old age, when this former theology student would accompany his family to church at Downe on Sundays but refuse to enter it himself. For those who find such sentiments too harsh or disturbing, here, as an alternative ending, is how Darwin himself ended the second edition of *The Origin* in 1860:

'Thus, from the war of nature, from famine and death, the most exalted object of which we are capable of conceiving, namely the production of higher animals directly follows. There is a grandeur in this view of life, with its several powers, having been originally breathed by the Creator into a few forms or into one; and that, while this planet has gone cycling on according to the fixed law of gravity, from so simple a beginning endless forms most beautiful and most wonderful have been, and are being, evolved.'

Further Reading

There are a very great number of books about Darwin and evolution. The following offer a broad introduction:

Attenborough, David. *Life on Earth: A Natural History*. Little, Brown, 1981.

Darwin, Charles. *The Origin of Species*. Signet Classic edition, 2003.

Darwin, Charles. *The Voyage of the Beagle*. Narrative Press edition, 2001.

Darwin, Charles. *The Descent of Man*. Princeton University Press edition, 1981.

Darwin, Charles. *Autobiography*. Norton edition, 1969.

Dawkins, Richard. *The Selfish Gene*. Oxford University Press, 2006.

Dawkins, Richard. *The Blind Watchmaker*. Penguin, 2006.

King-Hele, Desmond. *Doctor of Revolution: The Life and Genius of Erasmus Darwin*. Faber & Faber, 1977.

Pratchett, Terry, Jack Cohen, and Ian Stewart. *Darwin's Watch*. Ebury Publishing, 2006.

Strathern, Paul. *Crick, Watson, and DNA*. Anchor Books, 1999.

Watson, James D. *The Double Helix*. Simon & Schuster, 2001.

Acknowledgments

The late Eric Evans for arousing my interest in Charles and Erasmus Darwin, his son Dr Martin Evans for reading the text, and Professor Claudio Scazzocchio for a grounding in genetics.

Index